AF836590

MACHINE LEARNING FOR PLANETARY SCIENCE

MACHINE LEARNING FOR PLANETARY SCIENCE

Edited by

JOERN HELBERT

MARIO D'AMORE

MICHAEL AYE

HANNAH KERNER

ELSEVIER

Elsevier
Radarweg 29, PO Box 211, 1000 AE Amsterdam, Netherlands
The Boulevard, Langford Lane, Kidlington, Oxford OX5 1GB, United Kingdom
50 Hampshire Street, 5th Floor, Cambridge, MA 02139, United States

Copyright © 2022 Elsevier Inc. All rights reserved.

MATLAB® is a trademark of The MathWorks, Inc. and is used with permission.
The MathWorks does not warrant the accuracy of the text or exercises in this book.
This book's use or discussion of MATLAB® software or related products does not constitute endorsement or sponsorship by The MathWorks of a particular pedagogical approach or particular use of the MATLAB® software.

No part of this publication may be reproduced or transmitted in any form or by any means, electronic or mechanical, including photocopying, recording, or any information storage and retrieval system, without permission in writing from the publisher. Details on how to seek permission, further information about the Publisher's permissions policies and our arrangements with organizations such as the Copyright Clearance Center and the Copyright Licensing Agency, can be found at our website: www.elsevier.com/permissions.

This book and the individual contributions contained in it are protected under copyright by the Publisher (other than as may be noted herein).

Notices

Knowledge and best practice in this field are constantly changing. As new research and experience broaden our understanding, changes in research methods, professional practices, or medical treatment may become necessary.

Practitioners and researchers must always rely on their own experience and knowledge in evaluating and using any information, methods, compounds, or experiments described herein. In using such information or methods they should be mindful of their own safety and the safety of others, including parties for whom they have a professional responsibility.

To the fullest extent of the law, neither the Publisher nor the authors, contributors, or editors, assume any liability for any injury and/or damage to persons or property as a matter of products liability, negligence or otherwise, or from any use or operation of any methods, products, instructions, or ideas contained in the material herein.

ISBN: 978-0-12-818721-0

For information on all Elsevier publications
visit our website at https://www.elsevier.com/books-and-journals

Publisher: Candice Janco
Acquisitions Editor: Peter J Llewellyn
Editorial Project Manager: Andrea R. Dulberger
Production Project Manager: R. Vijay Bharath
Designer: Victoria Pearson

Typeset by VTeX

ELSEVIER Book Aid International
Working together to grow libraries in developing countries
www.elsevier.com • www.bookaid.org

Contents

Contributors

Rafael Alanis
Jet Propulsion Laboratory, California Institute of Technology, Pasadena, CA, United States

R.B. Anderson
U.S. Geological Survey, Astrogeology Science Center, Flagstaff, AZ, United States

I.P. Aneece
U.S. Geological Survey, Western Geographic Science Center, Flagstaff, AZ, United States

Erik Asphaug
Lunar and Planetary Laboratory, University of Arizona, Tucson, AZ, United States

Deegan Atha
Jet Propulsion Laboratory, California Institute of Technology, Pasadena, CA, United States

Michael Aye
Laboratory for Atmospheric and Space Physics, University of Colorado, Boulder, CO, United States

Saverio Cambioni
Department of Earth, Atmospheric & Planetary Sciences, Massachusetts Institute of Technology, Cambridge, MA, United States

Joseph Campbell
Carnegie Mellon University, Pittsburgh, PA, United States

Subhajit Chaudhury
The University of Tokyo, Department of Information and Communication Engineering, Tokyo, Japan

Flynn Chen
Yale University, New Haven, CT, United States
Jet Propulsion Laboratory, California Institute of Technology, Pasadena, CA, United States

Shreyansh Daftry

Jet Propulsion Laboratory, California Institute of Technology, Pasadena, CA, United States

Mario D'Amore

German Aerospace Center (DLR), Berlin, Germany

Annie Didier

Jet Propulsion Laboratory, California Institute of Technology, Pasadena, CA, United States

Gary Doran

Jet Propulsion Laboratory, California Institute of Technology, Pasadena, CA, United States

Roberto Furfaro

Systems and Industrial Engineering Department, University of Arizona, Tucson, AZ, United States

L.R. Gaddis

U.S. Geological Survey, Astrogeology Science Center, Flagstaff, AZ, United States

Universities Space Research Association, Lunar and Planetary Institute, Houston, TX, United States

Kevin Grimes

Jet Propulsion Laboratory, California Institute of Technology, Pasadena, CA, United States

Tatsuaki Hashimoto

The University of Tokyo, Department of Electrical Engineering and Information Systems, Tokyo, Japan

Jörn Helbert

German Aerospace Center (DLR), Berlin, Germany

Shoya Higa

Jet Propulsion Laboratory, California Institute of Technology, Pasadena, CA, United States

Tanvir Islam

Jet Propulsion Laboratory, California Institute of Technology, Pasadena, CA, United States

Yumi Iwashita

Jet Propulsion Laboratory, California Institute of Technology, Pasadena, CA, United States

Hannah Kerner

University of Maryland, College Park, MD, United States

Olivier Lamarre

University of Toronto, Toronto, ON, Canada

Jet Propulsion Laboratory, California Institute of Technology, Pasadena, CA, United States

Christopher Laporte

Jet Propulsion Laboratory, California Institute of Technology, Pasadena, CA, United States

J.R. Laura

U.S. Geological Survey, Astrogeology Science Center, Flagstaff, AZ, United States

Steven Lu

Jet Propulsion Laboratory, California Institute of Technology, Pasadena, CA, United States

Chris Mattman

Jet Propulsion Laboratory, California Institute of Technology, Pasadena, CA, United States

R. Michael Swan

Jet Propulsion Laboratory, California Institute of Technology, Pasadena, CA, United States

Masahiro Ono

Jet Propulsion Laboratory, California Institute of Technology, Pasadena, CA, United States

Kyohei Otsu

Jet Propulsion Laboratory, California Institute of Technology, Pasadena, CA, United States

Jordan Padams

Jet Propulsion Laboratory, California Institute of Technology, Pasadena, CA, United States

Sebastiano Padovan
German Aerospace Center (DLR), Berlin, Germany
EUMETSAT, Darmstadt, Germany
WGS, Darmstadt, Germany

Mike Paton
Jet Propulsion Laboratory, California Institute of Technology, Pasadena, CA, United States

Dicong Qiu
Carnegie Mellon University, Pittsburgh, PA, United States
Jet Propulsion Laboratory, California Institute of Technology, Pasadena, CA, United States

Brandon Rothrock
Jet Propulsion Laboratory, California Institute of Technology, Pasadena, CA, United States

Hiya Roy
The University of Tokyo, Department of Electrical Engineering and Information Systems, Tokyo, Japan

Sami Sahnoune
Jet Propulsion Laboratory, California Institute of Technology, Pasadena, CA, United States

Bhavin Shah
Jet Propulsion Laboratory, California Institute of Technology, Pasadena, CA, United States

Kathryn Stack
Jet Propulsion Laboratory, California Institute of Technology, Pasadena, CA, United States

Adam Stambouli
Jet Propulsion Laboratory, California Institute of Technology, Pasadena, CA, United States

Mark Strickland
Arizona State University, Tempe, AZ, United States

Vivian Sun
Jet Propulsion Laboratory, California Institute of Technology, Pasadena, CA, United States

Virisha Timmaraju
Jet Propulsion Laboratory, California Institute of Technology, Pasadena, CA, United States

Kiri L. Wagstaff
Jet Propulsion Laboratory, California Institute of Technology, Pasadena, CA, United States

Ingo P. Waldmann
University College London, London, United Kingdom

Toshihiko Yamasaki
The University of Tokyo, Department of Information and Communication Engineering, Tokyo, Japan

Foreword

Humans have used robots to explore other planets in our solar system since the beginning of planetary exploration. In the Apollo era, robotic spacecraft were sent to the Moon ahead of humans to collect critical information about lunar surface properties and landing sites. Today, scientists perform detailed scientific analyses and field geology on the surface of Mars using the Perseverance and Curiosity rovers, robotic vehicles equipped with cameras, spectrometers, and other instruments. Today, humans are actively exploring planets and other bodies in the solar system with more than 20 robotic spacecraft, with more to come in the next decade [1]. As humans send more robotic explorers into the solar system carrying increasingly sophisticated instruments, each observation contains even more information for scientists to analyze and increases the volume of archived data from planetary exploration missions. This is evidenced in Fig. 0.1, which shows the first close-up image of Mars ever taken in 1965 alongside an image of Mars taken nearly 50 years later in 2014. Analyzing the enormous volumes of data returned by past, present, and future planetary exploration missions will require scientists to adopt a different kind of robot—machine learning, a subfield of artificial intelligence which learns patterns, perceptions, and predictions from data in an automated way.

Machine learning is a topic that spans a broad range of methods, models, learning types, and machine behaviors. In planetary science, machine learning can be used to facilitate scientific discovery and analysis in multiple ways: by uncovering patterns or features of interest in large, complex datasets that are difficult for humans to analyze; by inspiring new hypotheses based on structure and patterns revealed in data; or by automating tedious or time-consuming tasks. The goal of this book is to provide a bridge between the communities of machine learning and planetary science to enable increased uptake of machine learning methods in planetary science, and improve the accessibility of planetary science data for the machine learning community. In the first chapter, we will cover the basics of machine learning, special considerations for applying machine learning to planetary science datasets, guidelines for implementing machine learning models, and resources for the reader to find a deeper understanding of machine learning methods if desired. In the second chapter, we will cover the types of data and challenges that are encountered in planetary science. The third chapter

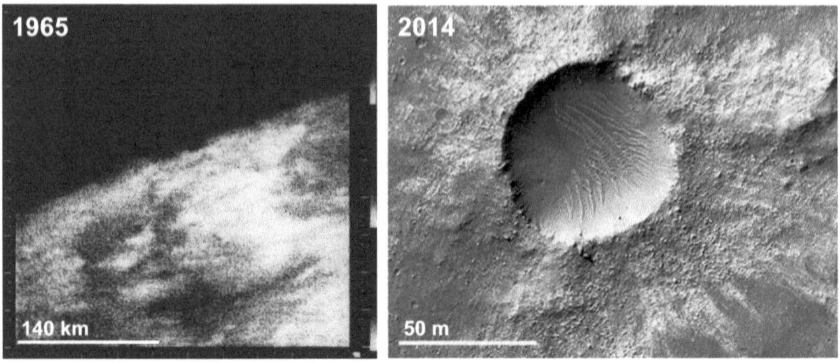

Figure 0.1 Left: first close-up image of Mars acquired by the Mariner 4 spacecraft on July 15, 1965. Credit: NASA/JPL-Caltech. Right: HiRISE image of impact crater on Mars acquired on January 9, 2014. Credit: NASA/JPL-Caltech/UA.

provides tutorials for accessing and preparing planetary science data sets for machine learning applications. The final chapter presents several case studies detailing how machine learning has been implemented for a variety of planetary science applications and data types.

References

[1] E. Lakdawalla, Where we are, The Planetary Report (2019) 22–23.

CHAPTER 1

Introduction to machine learning

Hannah Kerner[a], Joseph Campbell[b], and Mark Strickland[c]
[a]University of Maryland, College Park, MD, United States
[b]Carnegie Mellon University, Pittsburgh, PA, United States
[c]Arizona State University, Tempe, AZ, United States

1.1 Overview of machine learning methods

A widely used definition of *machine learning* (ML) comes from Carnegie Mellon University professor Tom Mitchell: "A computer program is said to learn from experience E with respect to some class of tasks T and performance measure P, if its performance at tasks in T, as measured by P, improves with experience E," or more simply, "machine learning is the study of computer algorithms that improve automatically through experience" [21]. In other words, machine learning involves using computer programs to learn some task based on experience (in the form of data or observations) where how well that task is learned is measured by some performance metric. For example, your email spam filter is a machine learning system: a computer program learns to classify emails as *spam* or *not spam* by observing millions of example emails that were known to be either *spam* or *not spam*, and how well it has learned to do this is measured by the percentage of correctly classified emails (the accuracy). Much of machine learning is considered in the optimization context, where the learning task is defined as learning the parameters θ (weights or coefficients) of a function $y=f(x,\theta)$ (where x is the input feature or features) by optimizing an objective function $C(y, y')$ where y is the correct value or label (*e.g.*, *spam* or *not spam*) and y' is the predicted value or label [21]. The objective function $C(y, y')$ (also called a loss function) compares y to y' and quantifies how close the prediction is to the correct value.

Machine learning describes a subset of methods within the broader field of artificial intelligence (AI). Computer vision is also a subarea of AI that describes methods for perception (including recognition and understanding) of images, videos, or other visual content. There is substantial overlap between computer vision and machine learning since machine learning methods (such as convolutional neural networks) are often employed for computer vision tasks (such as detecting craters in a satellite image of Mars),

Machine Learning for Planetary Science
https://doi.org/10.1016/B978-0-12-818721-0.00007-0
Copyright © 2022 Elsevier Inc.
All rights reserved.

though computer vision may also draw from nonML methods (e.g., SIFT, or scale-invariant feature transform [20]). The focus of this book is on the broader category of machine learning methods applied to planetary science, which includes but is not limited to computer vision tasks.

Machine learning methods are typically divided into the following subtypes:

(1) Supervised learning
(2) Unsupervised learning
(3) Semisupervised learning

Each of these methods represents a different type of learning based on what data is available as input to the learning model. In **supervised learning**, the data provided includes both a "feature vector" and a "label." The feature vector is a value (or set of values) intended to represent the observable feature(s), for a particular data sample. The label, in turn, is intended to represent a correct output for that sample. For example, one sample may consist of a feature vector in the form of an image (represented as a vector of pixel values) along with a label in the form of an integer that represents the category or "class" that describes the image. Fig. 1.1 illustrates an example input for classifying whether an image contains a crater or not: the input features are the pixels from the image and the class label is 1 if the image contains a crater and 0 if it does not. This is an example of a type of supervised learning called **classification**, in which the machine learning model maps input features to discrete output classes. The second type supervised learning is called **regression**, in which the model maps input features to a continuous output value (i.e., a real number). As an example, a model could be trained to predict the eccentricity of a crater in an image: the input features are the pixel values and the output is the real-valued eccentricity of the crater (Fig. 1.1).

In **unsupervised learning**, the input data is similar but no labels are available for training. In the following subsections, we will discuss methods for supervised and unsupervised learning in more detail. In **semisupervised learning**, models make use of both labeled and unlabeled data samples during training—in many applications there are far more unlabeled examples available than labeled examples. This is especially true in planetary science applications, partly because labels are more difficult to acquire than traditional machine learning applications because labeling may require domain knowledge or may not be clearly defined. For many planetary science applications, there may be very little data–labeled or unlabeled–requiring

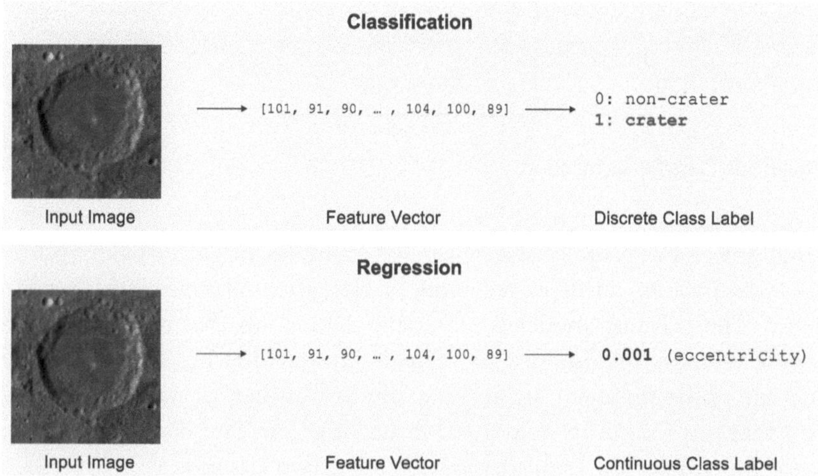

Figure 1.1 Example input and output pair for supervised classification (top) and regression (bottom).

specialized methods. Section 1.4 will cover methods for dealing with small labeled datasets, including semisupervised learning.

1.2 Supervised learning

Supervised learning refers to the set of machine learning techniques in which the model learns from input examples that contain an associated label for every example. There are two main categories of supervised learning techniques: classification, in which the model prediction is a discrete class or label, and regression, in which the model prediction is continuous (i.e., a real number).

1.2.1 Classification

Though supervised classification problems have inspired the majority of machine learning solutions, many real-world applications require significant effort to compile a training dataset since the data samples are usually labeled by humans. For example, in order to train a model to classify whether an image contains a crater or not (Fig. 1.1), one must first assemble a dataset containing hundreds or thousands of examples of images and an associated label of "crater" or "noncrater." Techniques for labeling planetary science datasets are discussed in more detail in Chapter 5 while techniques for

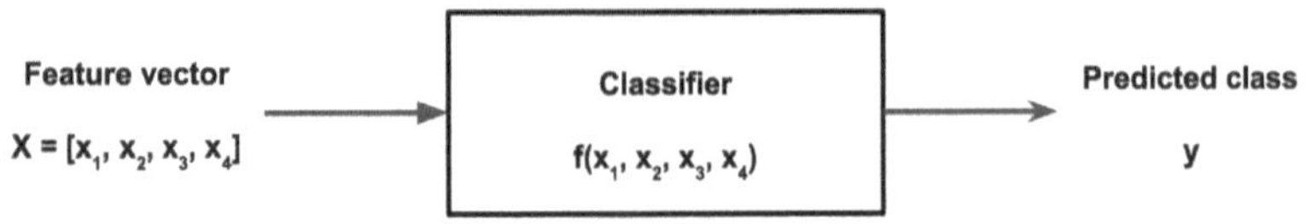

Figure 1.2 Classifier as function.

making use of a small number of labeled examples are discussed in Section 1.4. Classification involves predicting a class that corresponds to a given input. The learning model is essentially a function that maps the input feature values to the output class value, as illustrated in Fig. 1.2.

One of the simplest classifiers is a "linear classifier," in which the function that maps input to output is a linear function. For example, a simple linear function of the input that distinguishes two classes is:

$$y(x) = W^T\mathbf{x} + w_0 \tag{1.1}$$

where $\mathbf{x}$ is the input feature vector, W is called the weight matrix, and w_0 is called the bias. For this function, a "decision rule" can be used to distinguish the classes:

$$y = \begin{cases} C_1 & \text{if } y(x) \geq 0 \\ C_2 & \text{otherwise} \end{cases} \tag{1.2}$$

The "decision boundary" is therefore defined by the linear relation $y(x) = 0$, which is a hyperplane of dimension $N - 1$ where N is the dimensionality of the input features. One way to illustrate a linear classifier is shown in Fig. 1.3a for the case of a 2-dimensional feature vector. The blue (dark gray in print version) circles represent data samples from class C_1 while the red (mid gray in print version) squares are data samples from class C_2. The magenta (light gray in print version) line represents the decision boundary at $y(x) = 0$ that separates the classes.

Although the boundary shown in Fig. 1.3a is a simple line, the data samples that correspond to each class may not be linearly separable (e.g., Fig. 1.3b). In such cases, models that are capable of finding nonlinear decision boundaries (e.g., support vector machines and other kernel-based methods, or neural networks) or different feature extraction methods that enable the classes to be linearly separable are required.

A common obstacle in developing supervised classifiers is *overfitting*, which occurs when the classification model parameters are designed to

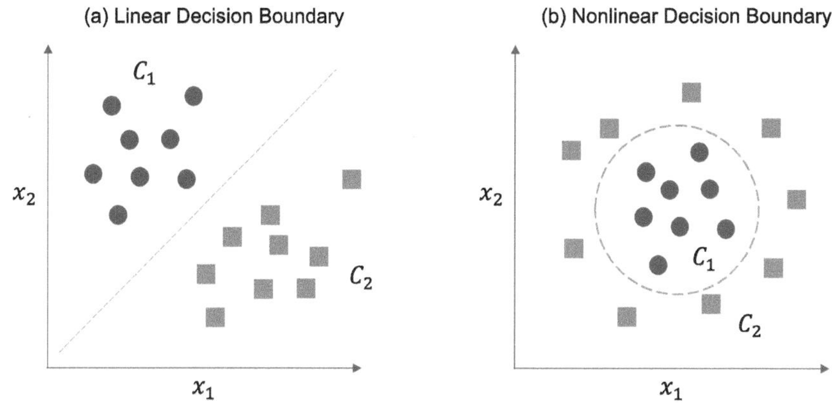

Figure 1.3 Linear decision boundary separating data samples from two classes.

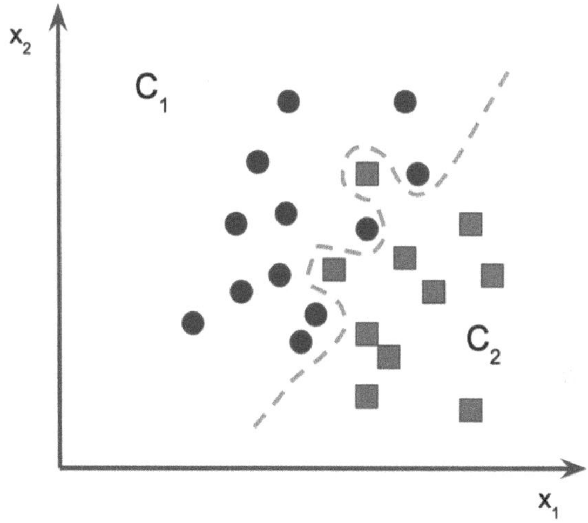

Figure 1.4 Example of overfitting.

perform particularly well for the training data, but the model performs relatively poorly when tested with data unseen during training (e.g., testing or validation data). Fig. 1.4 shows an example of a decision boundary for which overfitting has occurred. This problem can occur when the model parameters are specifically tailored to the training samples used rather than the underlying distribution of the samples. In this example, a higher-order polynomial was chosen to separate the classes C_1 and C_2 (magenta (light gray in print version) decision boundary), but the underlying distribution likely could be approximated using a simpler (e.g., linear) decision boundary and fewer model parameters.

1.2.2 Regression

Regression involves predicting a real number from a continuous range of possible values that corresponds to the input features. Regression can be considered analogous to curve-fitting: given a particular input, a regression model predicts a corresponding output that best fits the data used to train the model. One example of regression is the prediction of a weather-related parameter such as barometric pressure given a feature vector that includes values for temperature, humidity, and wind velocity. Another example of regression is prediction of stock or bond prices over time.

The regression model can take the form of a linear or nonlinear function of the input feature vector. The process of selecting a regression model includes choosing the order of the model. Following the curve-fitting analogy, selecting a regression model is analogous to choosing the order of a polynomial that fits the training data samples. If a polynomial of order one is chosen, the result is a linear regression model. If a quadratic or higher-order polynomial is chosen, the result is a nonlinear regression model.

For example, if we use linear regression to model a training dataset consisting of a number of data samples with corresponding continuous labels, we are essentially fitting the following linear model to the training data samples:

$$f(x) = w_1 x + w_0$$

where w_1 and w_0 are model parameters, also called regression coefficients. Here, x is the input feature vector and $y(x)$ is the output predicted value (or label). For a particular dataset with training data samples, the model parameters or regression coefficients can be determined using commonly-used methods, e.g., least squares, which attempt to find coefficients that minimize error between predicted value and the given label value.

1.3 Unsupervised learning

Unsupervised Learning refers to ML techniques in which the dataset consists of unlabeled data samples. An unsupervised learning model may be used to group similar data samples together (clustering) or map the input feature vector to an alternate (i.e., simplified or more compact) representation (dimensionality reduction). These applications are described in detail in this section. Other types of unsupervised learning applications include anomaly detection and probability density estimation.

In general, different types of learning models are used for unsupervised learning than those described in the previous section for supervised learning. This is because inputs for unsupervised learning models do not have corresponding labels, so the model must optimize a different target during learning—e.g., the likelihood of samples under a probability distribution. Unsupervised learning models can provide insights into the underlying structure of a dataset (e.g., identifying classes of minerals in a dataset of thermal infrared spectra), transform high-dimensional data samples into salient features that can be used as input for supervised techniques, and inspire new hypotheses based on patterns detected in planetary datasets.

1.3.1 Clustering

Clustering involves organizing data samples in a way that groups similar samples together. For example, a clustering algorithm could be used to group similar spectra observed by a spacecraft instrument and visualize the patterns of mineral formation on the surface. Chapter 7 describes an example of this for Mercury. These resulting groups or "clusters" often demonstrate patterns or commonalities that may not be immediately apparent without more extensive analysis of the samples. Because the number of groups is a finite number (often determined beforehand), the output of a clustering learning model is an identification of each group, along with an assignment of each data sample to one of the groups. Popular clustering methods include k-means clustering and spectral clustering.

In k-means clustering, a distance metric is chosen (e.g., Euclidean distance) and used to determine the distance between each pair of data samples. The Euclidean distances are used to partition the data samples into k clusters, in which the within-cluster sum of squared distances is minimized while also ensuring that each cluster contains at least one data sample. Although this type of clustering can be computationally expensive, approximate algorithms have been developed that iterate to local minima with lower computational cost.

Spectral clustering involves the generation of a similarity matrix relating each data sample to each other data sample (e.g., the inverse of the distance between samples). The eigenvalues and eigenvectors of the Laplacian of the similarity matrix are computed and then a clustering algorithm (e.g., K-means) is applied to the eigenvectors. Spectral clustering can be more computationally efficient than other clustering methods depending

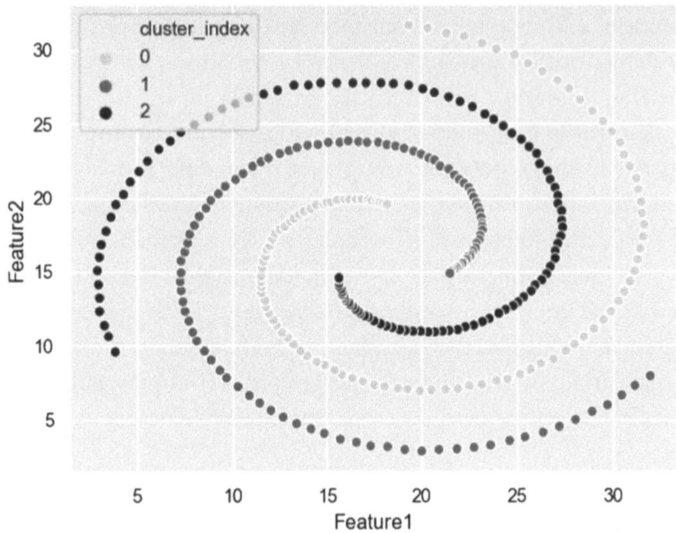

Figure 1.5 Example of Spectral Clustering.

on the sparseness of the similarity matrix. It can also result in improved cluster definitions for contiguous groups of data samples. See Fig. 1.5 for an example.

1.3.2 Dimensionality reduction

Many methods do not perform well when data samples are high-dimensional. This is referred to as the "curse of dimensionality." In these cases, methods to automatically reduce the dimensionality of data can be used to transform high-dimensional data samples into more compact representations that capture the most salient features in the data. This reduced representation can then be used as the input to supervised or unsupervised algorithms or to visualize high-dimensional data samples. At the core of dimensionality reduction is the manifold hypothesis, which is the theory that high-dimensional data actually lie on a lower-dimension manifold which can be represented by the sample's latent representation. For example, suppose we have a dataset containing 100×100-pixel images of faces, where each pixel can have an integer value between 0-255. The dimensionality of samples in this dataset is $100 \times 100 = 10{,}000$. However, the *intrinsic dimensionality* of these data is not 10,000. We know this because if we were to randomly sample from the distribution of all possible 100×100 pixel values, we would not get a face. Thus, the data must lie on a lower-dimensional manifold. Dimensionality reduction methods learn a mapping from high-dimensional data to lower-dimensional latent representations. Common algorithms for dimensionality reduction include principal component analysis (PCA), t-distributed Stochastic Neighbor Embedding (t-SNE), and autoencoder neural networks.

1.4 Semisupervised learning

The previous sections introduced both supervised and unsupervised learning paradigms, which exploit labeled data and unlabeled data respectively, in order to learn underlying latent information about the dataset. However, in planetary science it is often the case that labeled data is not plentiful enough to train a sufficiently accurate supervised learning model. This can be due to many factors, for example,

- Obtaining data for bodies other than Earth is extremely difficult and requires planning multiyear missions with tight constraints on what sensors can be used and where they can be employed.
- Creating accurate labels, such as image annotations for another planetary body, is time consuming and requires a high degree of expertise (making it difficult to utilize crowd-sourcing and citizen science approaches).
- Labels cannot be physically validated with on-site field surveys (barring exceptional circumstances), so low-confidence annotations must be treated with care.
- Communication is bandwidth- and power-constrained for bodies further away from Earth, limiting the amount of data that can be physically transmitted back. For example, the Cassini orbiter transmitted data at a rate between 14–166 kb/s once beyond the inner solar system.

While some of these issues might result in a small amount of labeled *and* unlabeled data, for now we concern ourselves with the more common case that we have a small set of labeled data and a larger set of unlabeled data. One way to approach this problem is to combine elements of both supervised and unsupervised learning in a complementary fashion, resulting in the aptly-named *semisupervised learning*.

Suppose we have a dataset consisting of surface images of a planetary body and we want to classify whether or not each image contains a recurring slope lineae. This is a binary classification problem and so a subset of our images have been annotated by experts as either $y = 1$ (contains an

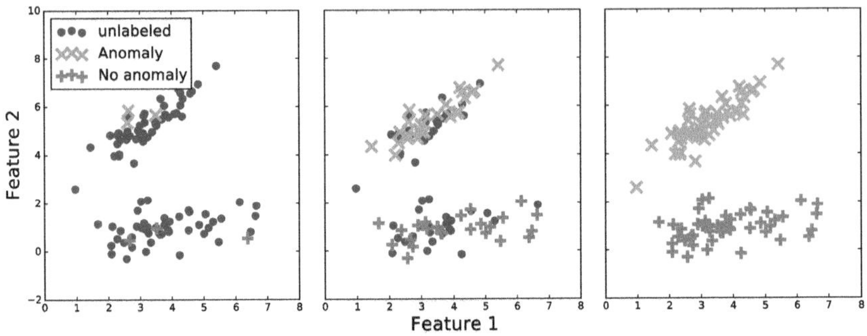

Figure 1.6 The labeled data points as produced by self-training: (left) initially, (middle) after half of the points have been processed, (right) after all of the points have been processed.

anomaly) or $y = 0$ (does not). However, given the time and expertise required to perform annotation, our labeled set of images is a small subset of the total image set available to us. Semisupervised learning provides an approach for exploiting what we know about the unlabeled data in order to use it to improve our classifier.

1.4.1 Self-training

In the simplest case, we train a classifier over the labeled set of images and use that to predict labels for the remaining unlabeled images. We then take the most confident predictions and add them to the labeled data set. The classifier is re-trained and we repeat this process until all images have been assigned labels. This process is visualized in Fig. 1.6. By repeatedly building up the labeled data set from its own predictions, the classifier is in a sense training itself – hence this is known as *self-training* [33].

The choice of which predicted label/image pairs to add to the training set in each iteration is typically based on confidence (thus requiring a probabilistic interpretation of the output label), although other criteria such as selection metrics based on distance have also been employed [27].

This is a simple, yet flexible method which allows the user to specify their own classification model. The caveat is that we assume the classifier's predictions are actually correct. If this is not the case and the classifier makes an incorrect prediction which is then added to the labeled data set, the algorithm can diverge from the true class labels by "doubling down" on its own mistakes, leading to worse classification results [35].

1.4.2 Self-training with Expectation Maximization

Alternatively, we can attempt to estimate the parameters for a *generative model* over the labeled and unlabeled data, which models the joint probability distribution $p(x, y)$ over input variables x and target variables y. This generative model is assumed to be a mixture distribution [25] or a classification model based on one, e.g., naive Bayes [23]. While it is straightforward to estimate the parameters for such a distribution with a maximum likelihood estimation approach [5] given the labeled data set, this is only a noisy estimate of the true distribution. We would like to use the large amount of unlabeled data that may be available in addition to the labeled data to improve our estimate of the model parameters, however, the unlabeled data on its own only allows us to estimate the marginal probability $p(x)$ over *all* class labels. In order to integrate the unlabeled data into our estimate of the joint distribution $p(x, y)$, we adopt an iterative two-step process in which we 1) estimate class label probabilities for each unlabeled sample given our current best estimate of the distribution parameters, and 2) improve our estimate of the distribution parameters given the new labels. This process repeats until the parameters converge and is known as Expectation Maximization [10], for which we treat the unknown labels as latent variables that are to be estimated along with the parameters.

As in regular self-training, the EM-based version iteratively trains itself over time; the difference lies in how unlabeled data is used in training. In self-training only a subset of the predictions (the most confident) are incorporated into the labeled data set at each step. However, in self-training with EM, all predictions are weighted by their expectation and immediately incorporated into training. Thus, self-training can be thought of as an *incremental* algorithm and self-training with EM as an *iterative* algorithm. While EM is well-studied and has been applied to many different applications, it requires strong assumptions to be made about the underlying distribution of the data that cause performance degradation when violated; namely, that it is generated by the proposed mixture model and that there is a relationship between the model components and the data classes [22].

1.4.3 Cotraining

In a slightly more complex strategy, we can instead opt to subdivide the data into two distinct feature sets in a procedure called *cotraining* [6]. Intuitively, these feature sets represent two different ways – or *views* – of looking at the same data. Consider the anomaly detection classification scenario from before and suppose that we have two input features associated with our images: pixel color and topographic elevation denoted as x_c and x_e respectively. One way of looking at the data is that pixel color alone is indicative of whether there is an anomaly or not, and so we can train a classifier C_c to predict if there is an anomaly based solely on x_c with the labeled data set. A different, complementary view is that topographic elevation is sufficient to indicate the presence of an anomaly, and we train a second classifier C_e over this view using the input feature x_e. The fundamental idea behind cotraining is that each classifier can identify complementary information about the unlabeled data set that can be used to build up the other, in a fashion similar to self-training. The color classifier C_c is used to predict labels for the unlabeled data set and a subset of these predictions is selected—often the most confident predictions—and added to the labeled data set. Similarly, C_e predicts labels for the unlabeled instances and also selects a subset to be added. Then both classifiers are re-trained over the new labeled data and this iterative process repeats. Without diving into formal theory, the reason that this works is that both classifiers must agree on the predicted labels for the unlabeled data. When considering the hypothesis space of all models, the net effect is that we prune models that do not agree, thus reducing the overall size of the space [35]. As in self-training, this method allows flexibility in the choice of the classification model, however, there are assumptions which must hold in order to provably learn [22]: a) each view must be sufficient to predict the class label on its own *and* the two views must agree over most of the labels, and b) the views must be conditionally independent given the class label.

1.5 Active learning

Semisupervised learning seeks to exploit what we know about the unlabeled data in order to improve our model performance from a limited set of labeled samples. There is an opposite, yet complementary approach known as *active learning* which seeks to exploit what we *don't* know about the unlabeled data [29]. A subset of unlabeled samples is queried for the class label; intuitively, if we assume that the act of labeling samples is a time-consuming task then if we can carefully select which samples would be most informative then we can limit the number of labeled samples that are required.

1.5.1 Uncertainty sampling

A simple strategy for selecting which samples to label is to query the labels for the instances in which the classifier is least confident about its prediction. This strategy is referred to as *uncertainty sampling* [19]. This is an iterative method in which a classifier is first trained on the labeled data set, predicts labels for the unlabeled samples, then queries labels for the samples in which it is least confident (e.g., whose predictions are closest to the decision boundary). These newly labeled samples are then labeled and added to the labeled data set and the process is repeated until all unlabeled samples can be confidently predicted. Uncertainty sampling can be seen as the active learning counterpart to self-training—in uncertainty sampling we are querying labels for the least confident predictions rather than adding the most confident as in self-training. The choice of classification model is constrained to those that are capable of providing a probabilistic output or a similar measure of uncertainty.

1.5.2 Query-by-committee

A more complex approach called *query-by-committee* [30] is to partition the model hypothesis space by training multiple classification models over the labeled data set. Each classifier then predicts labels for the unlabeled samples and we query labels for the samples on which the predictions do not agree. Two assumptions are made by this method: a) that we can partition the hypothesis space with multiple classification models, and b) that these models disagree on a subset of unlabeled samples. As in cotraining, we seek to reduce the size of the hypothesis space to narrow down the list of candidate models. However, while cotraining gradually builds up the models such that they always agree, here we achieve the same effect by querying for labels in regions where they disagree.

1.6 Popular machine learning methods

While there are myriad models and algorithms to choose from when implementing the approaches discussed in this chapter, there is a much smaller subset of more commonly used algorithms. We will summarize common methods in this section to provide a starting point for those looking to apply machine learning and data science techniques for planetary science. We illustrate these methods using an example dataset of HiRISE grayscale orbital images of Mars labeled as crater, bright dune, dark dune, slope streak, impact ejecta, swiss cheese, spider, or other [12]. This dataset is also used in Chapter 5 and Wagstaff et al. [32]—see Fig. 5.7 in Chapter 5 for an example image in each class. We excluded the 'other' class to simplify our visualizations since this class has substantially higher intra-class variance than other classes. The python code used to create the visualizations in this chapter is available at https://github.com/hannah-rae/planetary-ml-book.

In general, the choice of which model to use will depend on a variety of factors specific to the application, such as number of training labels, dimensionality of data samples, computational speed or storage requirements, or interpretability. Scikit-learn, a popular machine learning library for python, provides a flowchart to help users choose the right algorithm for their application: https://scikit-learn.org/stable/tutorial/machine_learning_map/index.html. Readers may also find Raghu & Schmidt [24] helpful for understanding if a neural network approach, and which type of neural network, is best suited for a scientific application.

1.6.1 Principal component analysis

Principal component analysis (PCA) is a dimensionality reduction method used to project data to a lower-dimensional space. PCA is widely used in planetary science—for example, Chapter 8 uses PCA for exploratory data analysis of hyperspectral image observations of Saturn from the Cassini mission. PCA defines a linear projection of the data onto a principal subspace that retains maximal variance in the data. The principal components are the eigenvectors of the data covariance matrix, which can be computed using singular value decomposition (SVD). PCA is commonly used to reduce the dimensionality of high-dimensional data sets for feature extraction, visualization, or other purposes. The number of principal components to retain (often referred to as the variable k) is a hyperparameter that can be chosen or tuned. A common approach to choosing k is to determine when the fraction of total variance explained by the k components starts to plateau for higher values of k. Fig. 1.7 shows a plot of the percentage of total variance explained by k principal components in the HiRISE dataset for $k \in [0, 20]$ and the dataset projected onto the 2-dimensional principal subspace.

1.6.2 K-means clustering

K-means is a popular unsupervised clustering algorithm that tries to assign data points to k clusters such that each sample belongs to the cluster whose mean is closest to the sample value. The standard algorithm begins

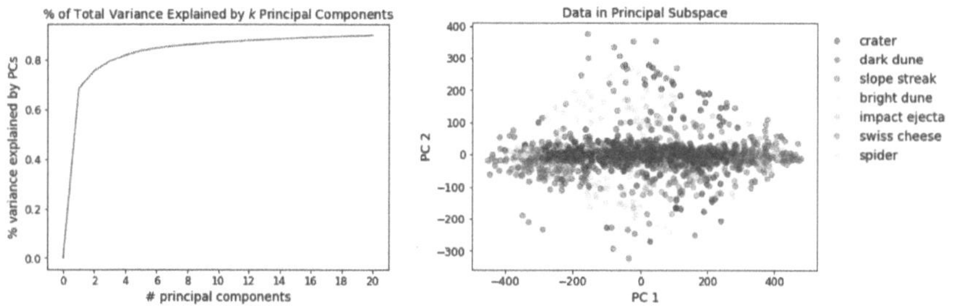

Figure 1.7 Left: percentage of total variance explained by k principal components for $k \in [0, 20]$. Right: HiRISE dataset projected onto principal subspace ($k = 2$).

by randomly initializing k cluster centroids and iteratively performing two steps: 1) assigning points to the centroid with the nearest cluster mean, and 2) re-computing each cluster mean (i.e., the mean of all samples assigned to the cluster). The algorithm converges and stops updating cluster assignments when the cluster assignment is no longer changing with each update. The Euclidean distance is often used as the distance metric for finding the nearest centroid. The number of clusters k is a tunable hyperparameter, which could be chosen based on domain knowledge (e.g., number of minerals likely to be present if clustering spectra) or such that within-cluster variance is minimized. Because the Euclidean distance tends to be an ineffective metric for high-dimensional data sets, K-means is frequently combined with PCA or other dimensionality reduction algorithms so that clustering is performed in the low-dimensional space. Chapter 7 uses K-means clustering with dimensionality reduction and manifold learning techniques to map similar classes of surface composition on the surface of Mercury using data from the MESSENGER mission. Fig. 1.8 shows the clusters identified using the k-means algorithm with $k = 7$. While some clusters appear to correlate with a class label when comparing to Fig. 1.7 (e.g., dark dune and cluster 2), overall the cluster assignments do not map directly to the class labels, as is often the case for high-dimensional (e.g., image) data sets.

1.6.3 Support vector machines

Support vector machines (SVMs) [5] are a supervised learning method that finds the hyperplane (or set of hyperplanes) in the n-dimensional feature space (where n is the number of input features) which maximizes the distance to the nearest training samples from each class. Maximizing this

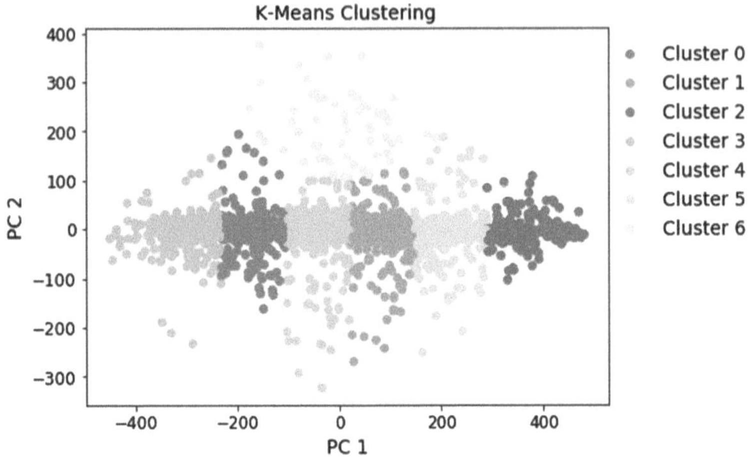

Figure 1.8 K-means clusters of PCA-reduced HiRISE data set.

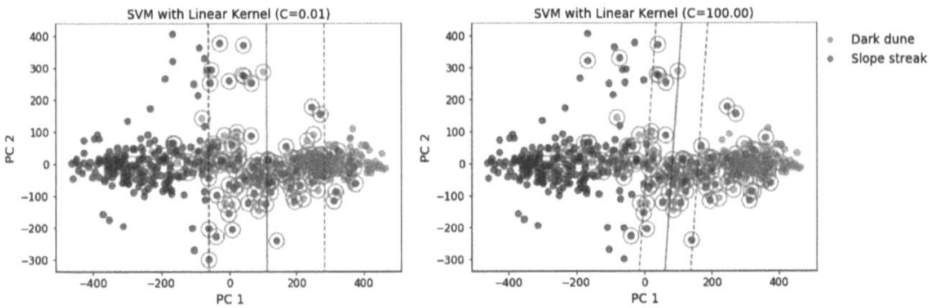

Figure 1.9 SVM decision boundary (solid gray line) for two classes in the HiRISE data set with two different values of hyperparameter *C*. The dashed lines show the margin and the circled points are the support vectors.

margin ensures that the decision boundary will be as generalizable as possible to new, unseen data points. SVMs can be used for both classification and regression. The main hyperparameter to choose is the regularization parameter C, which represents the strength of the penalty incurred during training for allowing samples to be closer to the margin boundary (since a perfect decision boundary is not attainable for most problems). SVM also uses a kernel function K to map samples to a higher dimensional space (this is referred to as the "kernel trick" [5]); the radial basis function (RBF) is a commonly chosen kernel function. Fig. 1.9 visualizes the decision boundary found by SVM on the HiRISE data set with PCA using a linear kernel (we reduced the data set to two classes to simplify the visualization). The

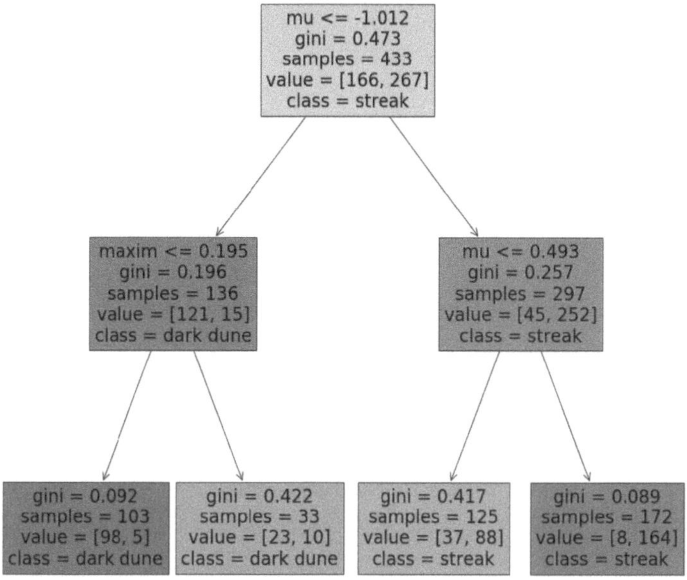

Figure 1.10 Decision tree with maximum depth of 2 learned from two classes of the HiRISE data set. Box colors represent the majority class and the Gini impurity (lighter colors represent higher impurity) for each node.

two plots illustrate the effect of smaller ($C = 0.01$) and larger ($C = 100$) values of C on the decision boundary.

1.6.4 Decision trees and random forests

Decision trees [8] are a supervised learning method used for both classification and regression that predict a target variable by learning a set of rules that progressively narrow down the decision space. These rules are inferred from the training data and have a simple "if-then-else" structure. For classification, the model learns rules that split the data in such a way that the "impurity" (e.g., measured by Gini impurity) of training samples is minimized. The impurity refers to the degree to which each branch of a split contains more than one class. For regression, a scalar loss metric such as Mean Squared Error or Mean Absolute Error is minimized. The maximum depth of the tree is a hyperparameter that can be tuned: the higher the depth of the tree, the more complex and susceptible to overfitting the model is. Fig. 1.10 shows the decision tree learned from the HiRISE data set maximum depth set to 3. We used only two classes and statistics features (mean, standard deviation, minimum, maximum, skew, and kurtosis) instead of PCA to simplify the visualization and provide interpretability.

The random forest method [7] is an extension of decision trees meant to combat overfitting that is typically experienced with decision trees in practice. A random forest is essentially an ensemble of decision trees in which each tree is learned from a random subset of samples from the training set. Each sample is drawn with replacement. This is referred to as "bootstrapping." Since the parameters of each decision tree are very sensitive to the training samples, sampling with replacement allows many different samples to be drawn from the same training data set to create many different trees that are ensembled in the final prediction. The final random forest prediction is obtained by averaging or taking a vote of the predictions from all of the trees. Tunable hyperparameters include the number of trees, maximum tree depth, and number of samples drawn for each tree. Random forests often show good performance in practice and are commonly used in scientific applications such as remote sensing [3].

1.6.5 Neural networks

A neural network consists of layers of composable nonlinear modules ("neurons") that each transform the data representation from the previous layer into a new representation that is more abstract than at the previous layer. Every neural network contains an input layer, one or more hidden layers, and an output layer; the network is called "deep" if it contains two or more hidden layers. Inspired by neurons firing in human brains, the output from each neuron is also called the "activation" and computes the following function:

$$a = h(\boldsymbol{W}^T \boldsymbol{X} + \boldsymbol{b}) \tag{1.3}$$

where $h(\cdot)$ is a nonlinear function (called the activation function), $\boldsymbol{X}$ is a tensor containing outputs from neurons in the previous layer (or the network input, in the case of the first hidden layer), $\boldsymbol{W}$ is a tensor containing weights w_{ij}^k representing the kth weight connecting layer i to layer j (purple arrows in Fig. 1.11), and $\boldsymbol{b}$ is a tensor of biases. During training, the parameters $\boldsymbol{W}$ and $\boldsymbol{b}$ are continually updated using a standard procedure called backpropagation such that a cost function $C(\cdot)$, also called the loss function, is minimized.

By learning many layers of hierarchical representations from input data, deep neural networks are able to discover complex structures in high-dimensional data (*e.g.*, images) that ignore irrelevant variations while fo-

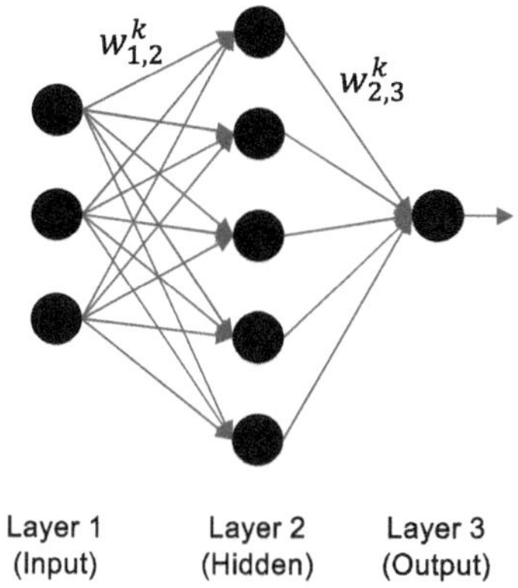

Figure 1.11 Example feed-forward neural network with neurons shown in black and weights in purple.

cusing on subtle but important variations [18]. The primary difference between deep learning or neural network techniques compared to traditional machine learning approaches is that traditional methods typically require manual specification of how to transform raw data (*e.g.*, pixel values) into feature representations suitable for the learning algorithm, requiring significant engineering effort and usually domain expertise. In neural networks, the representation does not need to be designed by human engineers; instead, it is learned automatically from the data using a general-purpose procedure. Another important difference is that traditional algorithms, such as support vector machines and random forests, typically require 1D vectors as input, which requires that multidimensional samples (such as multispectral images) be flattened into 1D vectors or transformed into a feature vector, thus removing spatial or local context. On the other hand, methods that can ingest multidimensional tensors and preserve the spatial topology of inputs such as convolutional neural networks can learn 2D or higher dimensional patterns in the data.

The primary disadvantage of deep learning compared to traditional machine learning techniques is that they require large labeled datasets (thousands or even millions of examples, depending on the architecture) for tuning the network parameters and their complex architectures can make it difficult to interpret the representations and decision functions learned by the network. A common approach for applications with smaller data sets is to pretrain a neural network on a large generic data set such as ImageNet [11] and then fine-tune the network on the smaller domain-specific data

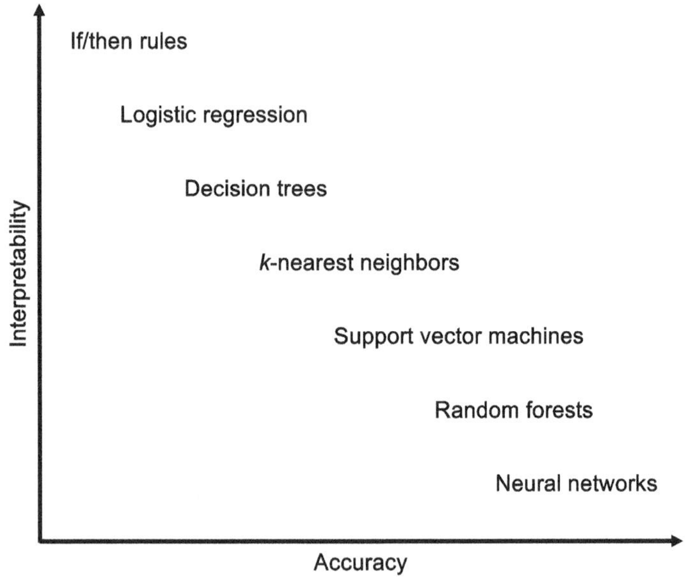

Figure 1.12 Relative rankings of popular classification models on the basis of accuracy and interpretability.

set. Fine-tuning consists of initializing a network with weights from the pretraining data set and further tuning the weights, often in the final or penultimate layers only, based on a different data set with different output categories. The effectiveness of fine-tuning relies on the assumption shown in prior work that the representations learned by a neural network trained on a large dataset are generic and can be useful for classification tasks that may be very different from the original task [26]. Chapter 5 provides a tutorial for fine-tuning a neural network on the HiRISE data set that was pretrained on the ImageNet data set.

Another important limitation of neural networks is that they are less interpretable than traditional ML methods. Fig. 1.12 shows the relative trade-offs between popular classification methods on the axes of accuracy and interpretability. While neural networks tend to have higher accuracy than traditional methods, they are the least interpretable. Conversely, if/then rules are highly interpretable but not very accurate. These trade-offs are important to consider because in many scientific and real-world applications, it may be desirable to sacrifice some accuracy for more interpretability. Developing methods for interpreting neural networks is an active area of research that is summarized well by this tutorial: https://explainml-tutorial.github.io/.

Popular deep learning methods include feed-forward neural networks, recurrent neural networks, convolutional neural networks, autoencoder networks, and generative adversarial networks [13,14,18]—several of these are applied in the case study chapter of this book. Using these deep learning methods, scientists have achieved unprecedented results using deep learning in image processing in recent years, including high-resolution image generation [1,2,17], image-to-image [15,34] and sequence-to-sequence [9] translation, and visual question answering [28]. We refer the reader to LeCun et al. [18] for a general overview of deep learning, Raghu & Schmidt [24] for an overview of deep learning for scientific discovery, Karpatne et al. [16] and Bergen et al. [4] for surveys of deep learning in the geosciences, and Zhu et al. [36] for an overview of deep learning in remote sensing. Chapters 6, 8, 9, and 10 present case studies using a variety of deep learning methods for both classification and regression in planetary science applications.

1.7 Data set preparation

A typical machine learning workflow starts with a data preparation stage. In addition to dataset-specific preprocessing, e.g., removing clouds or data artifacts, coregistering observations, or extracting spectral indices or other features, three datasets are usually created from the available data: the training, validation, and test datasets. The training set is used for training the model, i.e., iteratively tuning model parameters or weights to optimize some cost or objective function. The validation set is used for selecting model *hyperparameters*. Hyperparameters are variables not tuned during training but that determine the configuration of the model and influence the learning process, such as the number of trees in a random forest. Finally, the test set is used for evaluating model performance on unseen data.

The examples in each of these datasets should be independent and nonoverlapping. If the source data set contains independent and identically-distributed examples, it is common to split the source data set into training, validation, and test sets using a 80%/10%/10%, 70%/15%/15%, or 60%/20%/20% split. Using a 80%/10%/10% split, a random set of 80% of the available examples will be assigned to the training set while 10% will be assigned to the validation and test sets each. However, in many real-world (including planetary science) data sets, there exist correlations between examples in the source data set that should be considered when creating these three data sets. This is important because correlations between examples in the training and validation or test sets can result in overestimation of model performance. This is especially important for geospatial data since, as stated succinctly by Tobler [31], "all things are related, but nearby things are more related than distant things." For example, in Chapter 5, the HiRISE data set contains 227 × 227-pixel image tiles cropped from HiRISE images with a much larger size and footprint. Since tiles cropped from the same HiRISE images would have higher correlation than tiles from different HiRISE images, Lu et al. divided the tiles into train, validation, and test sets in such a way that all landmarks that originated from the same HiRISE source image were confined to the same subset.

In some data sets, especially small data sets, model performance might be sensitive to the particular split of the data and thus the validation or test performance is not an accurate measure of the model performance across the entire data set. A popular technique to address this is k-fold cross validation. In k-fold cross validation, the data set is split into k folds or partitions each of size N/k where N is the number of samples in the data set. The model is then evaluated on each of the k folds, using the remaining folds as training data. Leave-one-out cross validation (LOOCV) is a special case of k-fold cross validation in which each fold contains only one example. Performance metrics are averaged across the k iterations.

References

[1] G. Antipov, M. Baccouche, J.-L. Dugelay, Face aging with conditional generative adversarial networks, eprint, arXiv:1702.01983, 2017.
[2] M. Arjovsky, S. Chintala, L. Bottou, Wasserstein GAN, eprint, arXiv:1701.07875, 2017.
[3] M. Belgiu, L. Drăguţ, Random forest in remote sensing: a review of applications and future directions, ISPRS Journal of Photogrammetry and Remote Sensing 114 (2016) 24–31.
[4] K.J. Bergen, P.A. Johnson, V. Maarten, G.C. Beroza, Machine learning for data-driven discovery in solid Earth geoscience, Science 363 (2019).
[5] C.M. Bishop, in: Pattern Recognition and Machine Learning, Springer, 2006, Chapter 7.
[6] A. Blum, T. Mitchell, Combining labeled and unlabeled data with co-training, in: Proceedings of the Eleventh Annual Conference on Computational Learning Theory, 1998, pp. 92–100.
[7] L. Breiman, Random forests, Machine Learning 45 (2001) 5–32.
[8] L. Breiman, J.H. Friedman, R.A. Olshen, C.J. Stone, Classification and regression trees. Belmont, CA: Wadsworth, International Group 432 (1984) 151–166.
[9] C. Chan, S. Ginosar, T. Zhou, A.A. Efros, Everybody dance now, in: Proceedings of the IEEE International Conference on Computer Vision, 2019, pp. 5933–5942.

[10] A.P. Dempster, N.M. Laird, D.B. Rubin, Maximum likelihood from incomplete data via the em algorithm, Journal of the Royal Statistical Society: Series B (Methodological) 39 (1977) 1–22.
[11] J. Deng, W. Dong, R. Socher, L.-J. Li, K. Li, L. Fei-Fei, ImageNet: a large-scale hierarchical image database, in: CVPR09, 2009.
[12] G. Doran, E. Dunkel, S. Lu, K. Wagstaff, Mars orbital image (HiRISE) labeled data set version 3.2, https://doi.org/10.5281/zenodo.4002935, 2020.
[13] I. Goodfellow, J. Pouget-Abadie, M. Mirza, B. Xu, D. Warde-Farley, S. Ozair, A. Courville, Y. Bengio, Generative adversarial nets, in: Advances in Neural Information Processing Systems (NIPS), 2014, pp. 2672–2680.
[14] G.E. Hinton, R.R. Salakhutdinov, Reducing the dimensionality of data with neural networks, Science (New York, N. Y.) 313 (2006) 504–507, https://doi.org/10.1126/science.1127647.
[15] P. Isola, J.-Y. Zhu, T. Zhou, A.A. Efros, Image-to-image translation with conditional adversarial networks, in: Proceedings of the IEEE Conference on Computer Vision and Pattern Recognition, 2017, pp. 1125–1134.
[16] A. Karpatne, I. Ebert-Uphoff, S. Ravela, H.A. Babaie, V. Kumar, Machine learning for the geosciences: challenges and opportunities, IEEE Transactions on Knowledge and Data Engineering 31 (2018) 1544–1554.
[17] T. Karras, S. Laine, T. Aila, A style-based generator architecture for generative adversarial networks, in: Proceedings of the IEEE Conference on Computer Vision and Pattern Recognition, 2019, pp. 4401–4410.
[18] Y. LeCun, Y. Bengio, G. Hinton, Deep learning, Nature 521 (2015) 436–444, https://doi.org/10.1038/nature14539.
[19] D.D. Lewis, W.A. Gale, A sequential algorithm for training text classifiers, in: SIGIR'94, Springer, 1994, pp. 3–12.
[20] D.G. Lowe, Distinctive image features from scale-invariant keypoints, International Journal of Computer Vision 60 (2004) 91–110.
[21] T. Mitchell, Machine Learning, 1st ed., McGraw Hill, 1997.
[22] K. Nigam, R. Ghani, Analyzing the effectiveness and applicability of co-training, in: Proceedings of the Ninth International Conference on Information and Knowledge Management, 2000, pp. 86–93.
[23] K. Nigam, A.K. McCallum, S. Thrun, T. Mitchell, Text classification from labeled and unlabeled documents using em, Machine Learning 39 (2000) 103–134.
[24] M. Raghu, E. Schmidt, A survey of deep learning for scientific discovery, arXiv preprint, arXiv:2003.11755, 2020.
[25] J. Ratsaby, S.S. Venkatesh, Learning from a mixture of labeled and unlabeled examples with parametric side information, in: Proceedings of the Eighth Annual Conference on Computational Learning Theory, 1995, pp. 412–417.
[26] A. Razavian, H. Azizpour, J. Sullivan, S. Carlsson, Cnn features off-the-shelf: an astounding baseline for recognition, in: Proceedings of the IEEE Conference on Computer Vision and Pattern Recognition Workshops, 2014, pp. 806–813.
[27] C. Rosenberg, M. Hebert, H. Schneiderman, Semi-supervised self-training of object detection models, in: WACV/MOTION, vol. 2, 2005.
[28] A. Santoro, D. Raposo, D.G.T. Barrett, M. Malinowski, R. Pascanu, P. Battaglia, T. Lillicrap, D. London, A simple neural network module for relational reasoning, in: Proceedings of the 30th International Conference on Neural Information Processing Systems, 2017, pp. 4967–4976.
[29] B. Settles, Active learning literature survey, Technical Report, University of Wisconsin-Madison, Department of Computer Sciences, 2009.
[30] H.S. Seung, M. Opper, H. Sompolinsky, Query by committee, in: Proceedings of the Fifth Annual Workshop on Computational Learning Theory, 1992, pp. 287–294.

[31] W.R. Tobler, A computer movie simulating urban growth in the Detroit region, Economic Geography 46 (1970) 234–240.
[32] K. Wagstaff, Y. Lu, A. Stanboli, K. Grimes, T. Gowda, J. Padams, Deep Mars: Cnn classification of Mars imagery for the pds imaging atlas, in: Proceedings of the AAAI Conference on Artificial Intelligence, vol. 32, 2018.
[33] D. Yarowsky, Unsupervised word sense disambiguation rivaling supervised methods, in: 33rd Annual Meeting of the Association for Computational Linguistics, 1995, pp. 189–196.
[34] J.-Y. Zhu, T. Park, P. Isola, A. Efros, Unpaired image-to-image translation using cycle-consistent adversarial networks, in: IEEE International Conference on Computer Vision (ICCV), 2017.
[35] X. Zhu, A.B. Goldberg, Introduction to semi-supervised learning, Synthesis Lectures on Artificial Intelligence and Machine Learning 3 (2009) 1–130.
[36] X.X. Zhu, D. Tuia, L. Mou, G.-S. Xia, L. Zhang, F. Xu, F. Fraundorfer, Deep learning in remote sensing: a comprehensive review and list of resources, IEEE Geoscience and Remote Sensing Magazine 5 (2017) 8–36.

CHAPTER 2

The new and unique challenges of planetary missions

Jörn Helbert and Mario D'Amore
German Aerospace Center (DLR), Berlin, Germany

2.1 Introduction

2.1.1 50 years of Mercury exploration

A good example of the trend to larger and more complex datasets is a comparison between three missions to Mercury: the NASA missions Mariner 10 and MESSENGER and the upcoming ESA-JAXA BepiColombo mission. Those three missions span a time period of almost 50 years encompassing most of planetary exploration by missions. At the same time Mercury has surprised planetary scientists with every mission.

Mariner 10 was the seventh successful launch in the Mariner series and the first spacecraft to visit Mercury. It was also the first spacecraft to use the gravitational pull of one planet (Venus) to reach another (Mercury), and the first spacecraft mission to visit two planets. The spacecraft flew by Mercury three times in a retrograde heliocentric orbit and returned images and data on the planet. Mariner 10 returned the first-ever close-up images of Venus and Mercury [1,2]. The primary scientific objectives of the mission were to measure Mercury's environment, atmosphere, surface, and body characteristics and to make similar investigations of Venus. That spacecraft was able to map between 40 and 50% of the planet due to the orientation of the flybys. This revealed a surface covered in craters that looked at first glance similar to the Moon. As the first major surprise of Mercury Mariner 10 revealed the presence of a significant internal magnetic field, measured constituents of the exosphere and confirmed Mercury's high uncompressed density.

The NASA MESSENGER mission [3] performed three flybys of Mercury before entering orbit around Mercury at 00:45 UTC on March 18, 2011. MESSENGER's primary mission began on April 4, 2011, followed by two extended missions starting on March 17, 2012. By March 6, 2013 the mission had imaged 100% of the surface [4]. The mission ended with a

Machine Learning for Planetary Science
https://doi.org/10.1016/B978-0-12-818721-0.00009-4
Copyright © 2022 Elsevier Inc.
All rights reserved.

crash of the spacecraft on Mercury on April 30, 2015. MESSENGER was the first mission to enter orbit around Mercury. MESSENGER revealed the surface to be surprisingly rich in volatile elements like sulfur [5], provided the first detailed views of the so-called hollow structures indicating currently ongoing surface modification [6], provided visual evidence of past volcanic activity on the surface of Mercury [7] and strong evidence for water ice deposit in the permanently shadowed craters close to the poles of Mercury [8].

BepiColombo is an interdisciplinary mission to explore Mercury carried out jointly between the European Space Agency and the Japanese Aerospace Exploration Agency [9]. It consists of two spacecraft, the ESA provided Mercury Planetary Orbiter (MPO) and the JAXA provided Mercury Magnetospheric Orbiter (MMO) which fly in a stack configuration to Mercury propelled by the Mercury Transfer Module. Upon arrival at Mercury the modules will separate and enter their respective science orbits. The mission launched successfully in October 2018. After 6 Mercury flybys (as well as one Earth and two Venus flybys) the mission will enter orbit around Mercury in December 2025.

The following table shows the payload of each of the missions. When possible an attempt was made to align similar types of instruments on each of the missions.

Obviously the instruments on the three missions spanning almost 50 years can not be compared 1-to-1. The comparison in Table 2.1 is based on similar functionality in the widest terms. Focusing on the first three lines, Table 2.1 highlights exactly the growing complexity that requires new methods for data analysis. On first glance the Television Photography on Mariner 10 [1], MDIS on MESSENGER [3] and the SIMBIO-SYS on BepiColombo [10] are similar instruments that returned (or will return) images of the surface. However looking in details we see the increase in spatial and spectral resolution as well as in complexity of the data acquisition.

The Television Photography experiment on the NASA Mariner 10 consisted of two spherical (150 mm diameter) Cassegrain telescopes with eight filters, each attached to a GEC 1 inch vidicon tube camera (TV) (1500 mm focal length and 0.5 degree field of view) for narrow-angle photography. An auxiliary optical system mounted on each camera provided wide-angle (62 mm focal length, 11 x 14 degree field of view) photography by moving a mirror on a filter wheel to a position in the optical path. The TV picture consisted of 700 scan lines with 832 picture elements

Table 2.1 Comparison of the payloads of the NASA missions Mariner 10 and MESSENGER and the ESA BepiColombo Planetary Orbiter based on their functionality.

Mariner 10	MESSENGER	BepiColombo
Television Photography	Mercury Dual Imaging System (MDIS)	Spectrometer AND Imagers for MPO BepiColombo Integrated Observatory SYStem (SIMBIO-SYS)
-	Mercury Atmospheric and Surface Composition Spectrometer (MASCS)	SIMBIO-SYS
Extreme Ultraviolet Spectrometer	MASCS	Probing of Hermean Exosphere by Ultraviolet Spectroscopy (PHEBUS)
Triaxial Fluxgate Magnetometer	Magnetometer (MAG)	MPO-MAGnetometer (MPO-MAG)
Celestial Mechanics and Radio Science	Radio Science (RS)	Mercury Orbiter Radio-science Experiment (MORE)
Scanning Electrostatic Analyzer and Electron Spectrometer		Search for Exospheric Refilling And Emitted Natural Abundances (SERENA)
Energetic Particles Experiment	Energetic Particle and Plasma Spectrometer (EPPS)	SERENA
-	Mercury Laser Altimeter (MLA)	BepiColombo LasEr Altimeter (BELA)
Two-Channel Infrared Radiometer	-	MErcury Radiometer and Thermal infared Imaging Spectrometer (MERTIS)
-	Gamma-Ray Spectrometer (GRS)	Mercury Gamma ray and Neutron Spectrometer (MGNS)
-	Neutron Spectrometer (NS)	MGNS
-	X-Ray Spectrometer (XRS)	Mercury Imaging X-ray Spectrometer (MIXS)
-	-	Italian Spring Accelerometer (ISA)
-	-	Solar Intensity X-ray and particles Spectrometer (SIXS)

per line, which were digitally coded into eight-bit words for transmission. Exposure time ranged from 3 ms to 12 s, and each camera took a picture every 42 s. The average bit rate from this experiment was 22 kbps. The system had eight filter wheel positions: (1) wide-angle image relay mirror; (2) blue bandpass; (3) UV polarizing; (4) minus UV high pass; (5) clear; (6) UV bandpass; (7) defocusing lens (for calibration); and, (8) yellow bandpass. About 7000 photographs were obtained of Venus and Mercury, with a maximum resolution of 100 m for Mercury.

The Mercury Dual Imaging System (MDIS) on the NASA MESSENGER consists of a wide angle refractive optic imager with a 10.5 degree field of view and a narrow angle reflective optic imager with a field of view of 1.5 degrees. The wide angle imager has a modified achromatic Cooke-triplet lens which is 30 mm in diameter and a 12 position filter wheel with two clear filters centered at 750 nm, one with a 100 nm width, the other at 600 nm. The ten color filters are centered at 415 nm (40 nm width), 480 (30), 560 (10), 650 (10), 750 (10), 830 (10), 900 (10), 950 (20), 1000 (30), and 1020 (40). The light then passes through a small field flattening lens and strikes the 1024 x 1024 frame transfer CCD. The narrow angle imager uses an off-axis section of a Ritchie-Chretien reflective telescope to achieve a focal length of 550 mm with mirrors to correct spherical aberration and coma. The focal ratio is 18. A single band-limiting filter is used. As with the wide angle imager, a 1024 x 1024, 14 μm/pixel, frame transfer CCD is used. Each CCD records 12 bits/pixel and has manual and automatic exposure control over a range of 1 ms to 10 s.

The MASCS experiment consists of two instruments, a UV/Visible Spectrometer (UVVS) and a Visible/IR Spectrograph (VIRS) [11]. A baffled 250 mm Cassegrain f/5 telescope focuses light through a common boresight to both instruments. The UVVS consists of an Ebert-Fastie diffraction grating spectrometer. An 1800 groove/mm grating gives an average spectral resolution of 1.0 nm (0.5 nm in the far ultraviolet). The grating is rotated in 0.25 nm steps for scanning. Three photomultiplier tubes are situated behind separate slits, one covers the far ultraviolet (115-190 nm), one the middle ultraviolet (160-320 nm), and one the visible (250-600 nm). The VIRS is designed to measure surface reflectance in the 0.3-1.45 μm band with a spatial resolution of 100 m to 7.5 km. The field of view is 0.023 x 0.023 degrees. Light reaches the detector through a fused silica fiber optic bundle. A concave holographic diffraction grating with 120 lines/mm and a dichroic beam splitter which separates the visible (0.30-1.025 μm) and infrared (0.95-1.45 μm) parts of the spectrum are used to focus the spectra on two detectors. The visible detector is a 512 pixel silicon line array with an absorption filter in front of the long-wavelength half to eliminate the second order spectrum. The infrared detector is a 256 pixel InGaAs line array which does not require cooling. The Spectral resolution of the system is 4 nm and data is digitized to 12 bits.

The SIMBIO-SYS experiment on the ESA BepiColombo Mercury Planetary Orbiter [10] is an integrated suite for the imaging and spectroscopic investigation of the Hermean surface. It incorporates capabilities to perform medium space resolution global mapping in stereo and color imaging using two pan-chromatic and 3 broad-band filters, respectively, as well as high spatial resolution imaging in a pan-chromatic and 3 broad-band filters and imaging spectroscopy in the spectral range 400–2000 nm. It consists of three channels: the STereo imaging Channel (STC), with a broad spectral band in the 400-950 nm range and medium spatial resolution (at best 58 m/px), that will provide Digital Terrain Model of the entire surface of the planet with an accuracy better than 80 m; the High Resolution Imaging Channel (HRIC), with broad spectral bands in the 400-900 nm range and high spatial resolution (at best 6 m/px), that will provide high-resolution images of about 20% of the surface, and the Visible and near-Infrared Hyperspectral Imaging channel (VIHI), with high spectral resolution (6 nm at finest) in the 400-2000 nm range and spatial resolution reaching 120 m/px, it will provide global coverage at 480 m/px with the spectral information. The HRIC layout is based on a catadioptric optical design and consists of a modified Ritchey-Chretien (RC) configuration with a dedicated corrector. The instrument has a focal length of 800 mm and is equipped with a dioptric image corrector adapting the FoV to a 2048 x 2048 pixels detector with a pixel size of 10 μm. The STC camera consists of two subchannels with separate entrance optics and a common modified Schmidt telescope with a focal length of 95 mm. A correcting doublet lens replaces the classical Schmidt plate. The incoming light beams are focalized by the spherical mirrors on a 10 μm pixel size SiPIN hybrid CMOS detector. The STC detector allows snapshot image acquisition with a minimum integration time of 400 ns. Finally the VIHI channel concept is based on a collecting telescope and a diffraction-grating spectrometer ideally joined on the telescope focal plane, where the spectrometer entrance slit is located. The image of the slit is dispersed by the diffraction grating on a bi-dimensional detector. The 256×256 pixels detector array is built through hybridization of cadmium-mercury-telluride

(HgCdTe) sensors with the dedicated CMOS Read-Out Integrated Circuit (ROIC). The single pixel is 40×40 μm and the input circuit is a Capacitance Transimpedance Amplifier (CTIA) working in snapshot integration mode and allowing to readout while integrating.

The detailed look at these three imaging systems shows first of all the advances in available technology. However more importantly it shows that each mission has done very different design choices for their imaging systems. Typical machine learning datasets for image processing are based on consumer cameras which only differ in technical details. For imaging systems used on planetary missions this is not true. While some of the systems are framing cameras, mainly are line scanner or as in the case of MASCS even just point spectrometer. Color and/or spectral information is obtained in some cases with filters in some with dispersive elements, which influence for example if color information for a spot on the surface can be obtained at the same time and at the same geometry. Furthermore each of the imaging sensors provides its own set of advantages and challenges that can strongly influence the resulting data product to be analyzed.

Comparing the instrumentation we see as expected an improvement in spatial resolution. Mariner 10 had a maximum resolution of 100 m but most of the less than 50% of Mercury covered were at a resolution of a few kilometers/px. MESSENGER produced global color products at a resolution of 332 m/px and monochrome base maps at a resolution of 166 m/px. The highest spatial resolution MDIS could achieve was approximately 10 m. SYMBIO-SYS will image 20% of the planet at up 6 m/px, provide global imaging in stereo at up to 58 m/px and hyperspectral near-infrared data at 480 m/px. At the same time we move from 3 color filters on Mariner 10 to ten color filters for MDIS and a hyperspectral point spectrometer with MASCS on MESSENGER and hyperspectral imaging with the VIHI channel in SYMBIO-SYS on BepiColombo. Mariner 10 produced a total of approximately 7000 images of Venus and Mercury. MESSENGER produced a total of 694,232 images of Mercury. SYMBIO-SYS will produce several million images across its three subsystems. The increase in the number of data products alone requires new methods to process and analyze the data. Where 7000 images can be inspected visually this is impossible for almost 700,000 images or more. Here classification techniques as discussed in Chapter 4 become a key tool to preselect and analyze the data.

A second trend can be seen by the comparison between the three imaging system and that is the trend to increase complexity with the

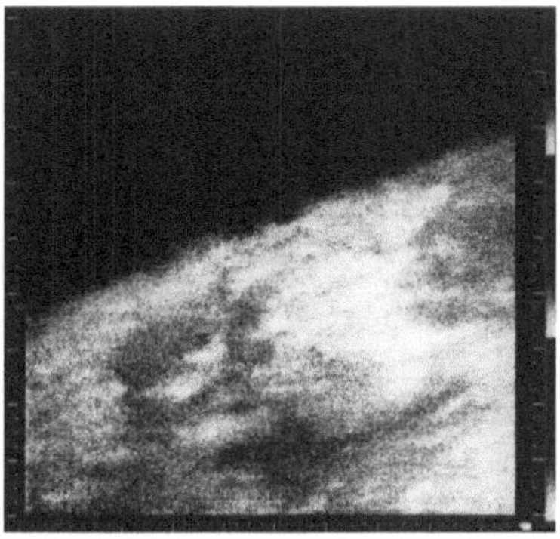

(a) Mariner 4 image, the first close-up image ever taken of Mars. This shows an area about 330 km across by 1200 km from limb to bottom of frame, centered at 37 N, 187 W. The area is near the boundary of Elysium Planitia to the west and Arcadia Planitia to the east. The hazy area barely visible above the limb on the left side of the image may be clouds. The resolution of this image is roughly 5 km and north is up. (NSSDC Data Set ID (Photo): 64-0771-01A)

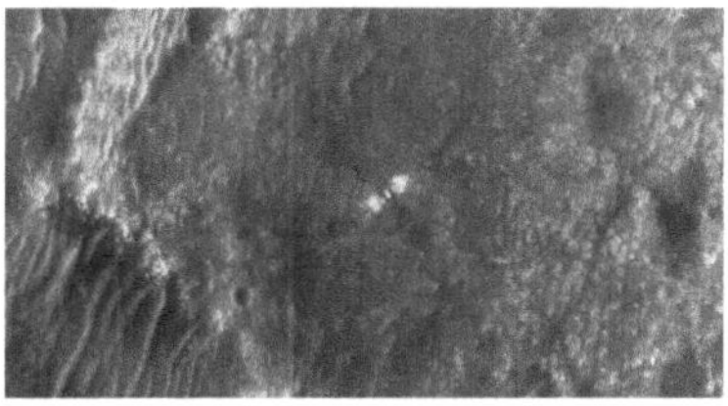

(b) The NASA Perseverance rover can be seen in this enhanced HiRISE color image at its landing site, six days after touchdown, doing system checks. The site appears to be covered in loose dark material with brighter material underneath. You can see the two bright zones to the sides of the rover that have been scoured clear by the descent stage rockets and the dark material appears to have been funneled outward both in front and behind the rover. (Credit: NASA/JPL/UArizona)

Figure 2.1 The evolution of Mars imaging from NASA Mariner 4 to HiRISE on NASA Mars Reconnaissance Orbiter.

instruments. The SYMBIO-SYS suite combines (and exceeds) the functionalities in the MESSENGER MDIS and MASCS system [12]. More generally if we look at the whole payloads we see a stark increase just in the number of payload elements. However we don't just have more elements, we also have more complex elements. Where MESSENGER has a x-ray spectrometer [5], BepiColombo has an imaging x-ray spectrometer [13]. Where MESSENGER has the MASCS point spectrometer BepiColombo has three imaging spectrometers with VIHI, PHEBUS and MERTIS [14].

The Mercury mission datasets spanning over 5 decades are a good example the general trends we see in planetary exploration. For Mars the exploration started on July 15, 1965 with the NASA Mariner 4 mission capturing only 22 images of about 1% of Mars at a spatial resolution of 5 km/px in black and white. In contrast the HIRISE camera on the NASA Mars Reconnaissance Orbiter is capable of imaging the Perseverance rover on the surface of Mars in color at a resolution of 29.5 cm/pixel (Fig. 2.1). This continues for Mars with images (and compositional data) taken on the surface at resolutions better than a mm/px.

2.1.2 Challenges of large and complex data return

Machine learning methods are key to fully using the synergies of such complex and multidimensional datasets machine learning methods are the key. While each of the separate datasets reveals parts of the puzzle it is really when we combine them that we can obtain deeper insights. Of course this poses special challenges. Looking at BepiColombo we will have imaging data at a spatial resolution of up to 6 m, topography obtained by stereo-imaging and laser altimetry, near- and thermal infrared hyperspectral data at 500 m, temperature maps at 2 km and elemental composition at 1-10 km. Combining all these datasets in a meaningful way (Fig. 2.2) to provide a holistic view of Mercury requires data fusion and machine learning approaches in order to be able to focus on the data interpretation instead of being limited by the ability to process data.

While the increasingly large and complex dataset promise a rich science return they pose severe challenges on the mission design side. MESSENGER had a data return of 100 Gbits/year - equivalent to 3 bits/s. The BepiColombo mission will return 1550 Gbits/year - that is equivalent to 50 kbits/s. For comparison the 2G EDGE standard for mobile phones that is now mostly discontinued has a maximum data rate of 100 kbits/s. Getting the data from the spacecraft to the ground is increasingly a limiting factor. Currently the NASA Mars Reconnaissance Orbiter (MRO) has the highest downlink rate of any planetary mission with an averaged data rate between 0.5 and 4 Mbps depending on Earth-Mars distance. This can be seen as the upper limit for the achievable downlink rate in the foreseeable future. Most missions are limited by geometry and/or power limitations to much lower downlink rates as is the case for BepiColombo, where the size and construction of the high-gain antenna is limited by the high temperatures in the Mercury environment. A way to optimize the use of the downlink available is the preselection and preprocessing of data onboard the spacecraft. Automated feature technique algorithms as discussed in Chapter 4 would allow to downlink only data sets that fit a certain criterion only downlink data that shows features not encountered before. Following this approach further the data observed can directly influence the mission operation. The JAXA MMO of BepiColombo is using a set of event detection criteria to drive the spacecraft operation. The MMO Data Processing Unit (DPU) monitors the data provided by all the sensors. The DPU can trigger data downlink at higher telemetry rate autonomously. Depending on the mission phase this is triggered either by known events (e.g. crossing of magnetospheric boundaries) or by unknown events. This need for autonomy

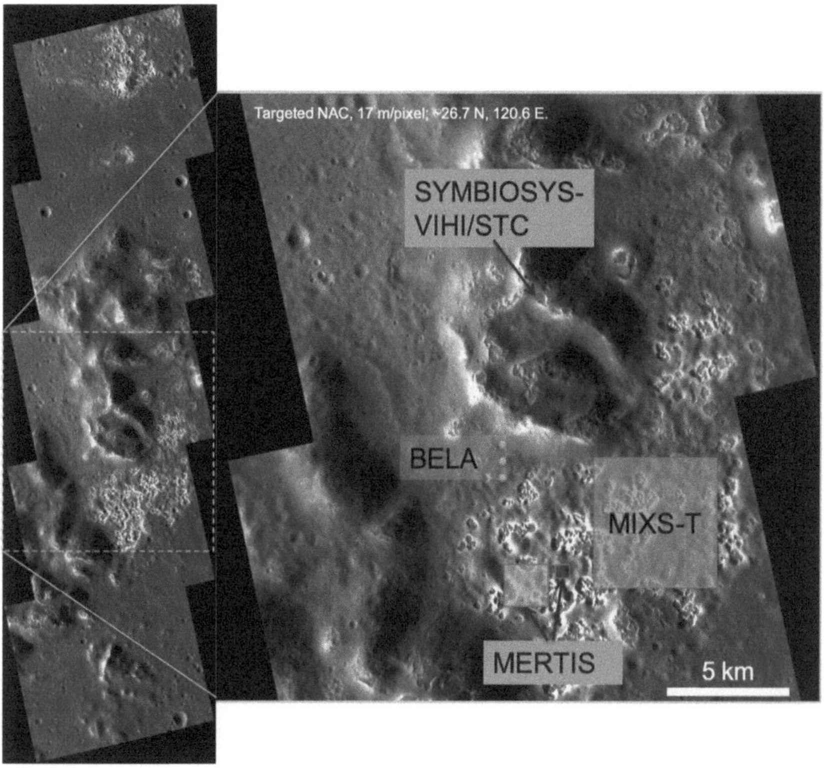

Figure 2.2 Raditladi peak ring on Mercury imaged by the NASA MESSENGER Narrow Angle Camera at 17 m/px. Overlaid are the footprints of payload elements on the ESA-JAXA BepiColombo mission, which will provide detailed spectral, elemental and geochemical data at resolutions between 5 and 1000 m/px. (Credit: NASA, John Hopkins Applied Physics Laboratory.)

in the data collection is driven by the constrained effective downlink rate of only 5 kbits/s for the MMO.

2.1.3 Facing the unknown

The formation of Mercury is one of the big questions in planetary science and additional data has made this question even more puzzling. How did the planet form so close to the Sun but yet retained large abundances of volatile species like sulfur on its surface? Why does the smallest of the terrestrial planets have an internal magnetic field like Earth when Mars and Venus have none? Why does Mercury have such a large core consisting most likely of iron but has a much lower iron oxide abundance on the surface than the Moon?

We currently have many ideas but no good answers to these questions. They highlight however the surprises that planetary scientists are facing when exploring planetary bodies. For example up until now we have no clear idea of the surface mineralogy of Mercury. So far the only mineral that has been identified tentatively was magnesium sulfide (MgS) [15]. There are no measured surface temperature maps available for Mercury and element composition maps are mainly available for the northern hemisphere.

This is a situation that this very different from typical application in remote sensing for Earth observation. For Earth observation we know first of all what to expect - which materials to expect and at what extent. Based on this we normally have clearly defined questions - where is a certain type of vegetation, where are deposits of certain ore, etc.? For planetary science we need to determine first the parameter space - what are actually the materials on the surface, as well as their distribution? Chapter 7 shows an example of this - trying to classify the surface units on Mercury is complicated by the fact that we have no a priori knowledge of how many different surface units exist and what are their compositions.

To make the situation even more interesting, we have to consider potential coupling between different observations that are unknown. To again use the Mercury example, the magnetic field that is protecting the surface from solar wind impingement can be compressed in certain solar wind conditions to expose surface regions to direct solar wind particle impact resulting in surface erosion and release of material into the exosphere. Studying this requires combining hyperspectral surface imaging, surface topography to consider shading, surface temperature map, time series of the magnetic field at the location of the spacecraft, particle flux observations from the spacecraft, solar activity observation either from the same or a different spacecraft as well as observations of the exosphere emission to trace surface release processes. As can be easily seen this combines datasets of different dimensions, data types, physical properties as well potentially even data from various spacecrafts. Machine learning approaches are ideally suited to explore hidden structures in this complex multidataset analysis approach.

2.1.4 Machine learning for planetary science

Machine learning methods are a key enabling technique for the analysis of data from planetary missions. This is driven by necessity due to the growing amount of data volume and complexity from planetary missions. At the same time these methods enable new scientific discoveries by combining complex and diverse datasets from a wide range of planetary instruments and missions. Planetary science requires a stronger focus on machine learning approaches that allow the exploration of unknown data than it is typically the case for Earth observation. The choice of methods depends on the properties of the data you have available, which varies for different planets as more missions are sent to them. We see this for Mars which has more missions than any other planet and thus more complete datasets. On the other hand for Venus we still have very limited data. However with the selection of three highly complementary missions to Venus that will return several terabytes of new data including the first global composition maps of the surface [16,17] this will soon change dramatically.

References

[1] J.A. Dunne, Mariner 10 Mercury encounter, Science 185 (1974) 141–142, https://doi.org/10.1126/science.185.4146.141, https://www.ncbi.nlm.nih.gov/pubmed/17810505.

[2] J.A. Dunne, Mariner 10 Venus encounter, Science 183 (1974) 1289–1291, https://doi.org/10.1126/science.183.4131.1289, https://www.ncbi.nlm.nih.gov/pubmed/17791368.

[3] S.C. Solomon, R.L. McNutt, R.E. Gold, D.L. Domingue, MESSENGER mission overview, Solar System Review 131 (2007) 3–39, https://doi.org/10.1007/s11214-007-9247-6.

[4] B.W. Denevi, N.L. Chabot, S.L. Murchie, K.J. Becker, D.T. Blewett, D.L. Domingue, C.M. Ernst, C.D. Hash, S.E. Hawkins, M.R. Keller, N.R. Laslo, H. Nair, M.S. Robinson, F.P. Seelos, G.K. Stephens, F.S. Turner, S.C. Solomon, Calibration, projection, and final image products of MESSENGER's Mercury dual imaging system, Solar System Review 214 (2018) 2, https://doi.org/10.1007/s11214-017-0440-y.

[5] L.R. Nittler, E.A. Frank, S.Z. Weider, E. Crapster-Pregont, A. Vorburger, R.D. Starr, S.C. Solomon, Global major-element maps of Mercury from four years of MESSENGER X-Ray Spectrometer observations, ICARUS 345 (2020) 113716, https://doi.org/10.1016/j.icarus.2020.113716, arXiv:2003.00650.

[6] Z. Xiao, R.G. Strom, D.T. Blewett, P.K. Byrne, S.C. Solomon, S.L. Murchie, A.L. Sprague, D.L. Domingue, J. Helbert, Dark spots on Mercury: a distinctive low-reflectance material and its relation to hollows, Journal of Geophysical Research (Planets) 118 (2013) 1752–1765, https://doi.org/10.1002/jgre.20115.

[7] J.L. Whitten, J.W. Head, B.W. Denevi, S.C. Solomon, Intercrater plains on Mercury: insights into unit definition, characterization, and origin from MESSENGER datasets, ICARUS 241 (2014) 97–113, https://doi.org/10.1016/j.icarus.2014.06.013.

[8] N.L. Chabot, C.M. Ernst, D.A. Paige, H. Nair, B.W. Denevi, D.T. Blewett, S.L. Murchie, A.N. Deutsch, J.W. Head, S.C. Solomon, Imaging Mercury's polar deposits during MESSENGER's low-altitude campaign, Geophysical Research Letter 43 (2016) 9461–9468, https://doi.org/10.1002/2016GL070403.

[9] J. Benkhoff, J. van Casteren, H. Hayakawa, M. Fujimoto, H. Laakso, M. Novara, P. Ferri, H.R. Middleton, R. Ziethe, BepiColombo—comprehensive exploration of Mercury: mission overview and science goals, Planetary and Space Science 58 (2010) 2–20, https://doi.org/10.1016/j.pss.2009.09.020.

[10] G. Cremonese, F. Capaccioni, M.T. Capria, A. Doressoundiram, P. Palumbo, M. Vincendon, M. Massironi, S. Debei, M. Zusi, F. Altieri, M. Amoroso, G. Aroldi, M. Baroni, A. Barucci, G. Bellucci, J. Benkhoff, S. Besse, C. Bettanini, M. Blecka, D. Borrelli, J.R. Brucato, C. Carli, V. Carlier, P. Cerroni, A. Cicchetti, L. Colangeli, M. Dami, V. Da Deppo, V. Della Corte, M.C. De Sanctis, S. Erard, F. Esposito, D. Fantinel, L. Ferranti, F. Ferri, I. FicaiÂ Veltroni, G. Filacchione, E. Flamini, G. Forlani, S. Fornasier, O. Forni, M. Fulchignoni, V. Galluzzi, K. Gwinner, W. Ip, L. Jorda, Y. Langevin, L. Lara, F. Leblanc, C. Leyrat, Y. Li, S. Marchi, L. Marinangeli, F. Marzari, E. MazzottaÂ Epifani, M. Mendillo, V. Mennella, R. Mugnuolo, K. Muinonen, G. Naletto, R. Noschese, E. Palomba, R. Paolinetti, D. Perna, G. Piccioni, R. Politi, F. Poulet, R. Ragazzoni, C. Re, M. Rossi, A. Rotundi, G. Salemi, M. Sgavetti, E. Simioni, N. Thomas, L. Tommasi, A. Turella, T. Van Hoolst, L. Wilson, F. Zambon, A. Aboudan, O. Barraud, N. Bott, P. Borin, G. Colombatti, M. ElÂ Yazidi, S. Ferrari, J. Flahaut, L. Giacomini, L. Guzzetta, A. Lucchetti, E. Martellato, M. Pajola, A. Slemer, G. Tognon, D. Turrini, SIMBIO-SYS: scientific cameras and spectrometer for the BepiColombo mission, Solar System Review 216 (2020) 75, https://doi.org/10.1007/s11214-020-00704-8.

[11] S.E. Hawkins, J.D. Boldt, E.H. Darlington, R. Espiritu, R.E. Gold, B. Gotwols, M.P. Grey, C.D. Hash, J.R. Hayes, S.E. Jaskulek, C.J. Kardian, M.R. Keller, E.R. Malaret, S.L. Murchie, P.K. Murphy, K. Peacock, L.M. Prockter, R.A. Reiter, M.S. Robinson, E.D. Schaefer, R.G. Shelton, R.E. Sterner, H.W. Taylor, T.R. Watters, B.D. Williams, The Mercury dual imaging system on the MESSENGER spacecraft, Solar System Review 131 (2007) 247–338, https://doi.org/10.1007/s11214-007-9266-3.

[12] W.E. McClintock, M.R. Lankton, The Mercury atmospheric and surface composition spectrometer for the MESSENGER mission, Solar System Review 131 (2007) 481–521, https://doi.org/10.1007/s11214-007-9264-5.

[13] E.J. Bunce, A. Martindale, S. Lindsay, K. Muinonen, D.A. Rothery, J. Pearson, I. McDonnell, C. Thomas, J. Thornhill, T. Tikkanen, C. Feldman, J. Huovelin, S. Korpela, E. Esko, A. Lehtolainen, J. Treis, P. Majewski, M. Hilchenbach, T. Väisänen, A. Luttinen, T. Kohout, A. Penttilä, J. Bridges, K.H. Joy, M.A. Alcacera-Gil, G. Alibert, M. Anand, N. Bannister, C. Barcelo-Garcia, C. Bicknell, O. Blake, P. Bland, G. Butcher, A. Cheney, U. Christensen, T. Crawford, I.A. Crawford, K. Dennerl, M. Dougherty, P. Drumm, R. Fairbend, M. Genzer, M. Grande, G.P. Hall, R. Hodnett, P. Houghton, S. Imber, E. Kallio, M.L. Lara, A. Balado Margeli, M.J. Mas-Hesse, S. Maurice, S. Milan, P. Millington-Hotze, S. Nenonen, L. Nittler, T. Okada, J. Ormö, J. Perez-Mercader, R. Poyner, E. Robert, D. Ross, M. Pajas-Sanz, E. Schyns, J. Seguy, L. Strüder, N. Vaudon, J. Viceira-Martín, H. Williams, D. Willingale, T. Yeoman, The BepiColombo Mercury imaging X-ray spectrometer: science goals, instrument performance and operations, Solar System Review 216 (2020) 126, https://doi.org/10.1007/s11214-020-00750-2.

[14] H. Hiesinger, J. Helbert, G. Alemanno, K.E. Bauch, M. D'Amore, A. Maturilli, A. Morlok, M.P. Reitze, C. Stangarone, A.N. Stojic, I. Varatharajan, I. Weber, Mertis Co-I Team, Studying the composition and mineralogy of the Hermean surface with the Mercury radiometer and thermal infrared spectrometer (MERTIS) for the BepiColombo mission: an update, Solar System Review 216 (2020) 110, https://doi.org/10.1007/s11214-020-00732-4.

[15] F. Vilas, D.L. Domingue, J. Helbert, M. D'Amore, A. Maturilli, R.L. Klima, K.R. Stockstill-Cahill, S.L. Murchie, N.R. Izenberg, D.T. Blewett, W.M. Vaughan, J.W. Head, Mineralogical indicators of Mercury's hollows composition in MESSENGER color observations, Geophysical Research Letter 43 (2016) 1450–1456, https://doi.org/10.1002/2015GL067515.

[16] J. Helbert, M.D. Dyar, D. Kappel, A. Maturilli, N. Mueller, Importance of orbital spectroscopy on Venus, Bulletin of the American Astronomical Society 53 (2021) 066, https://doi.org/10.3847/25c2cfeb.62e629cb.
[17] J. Helbert, A. Maturilli, M.D. Dyar, G. Alemanno, Deriving iron contents from past and future Venus surface spectra with new high-temperature laboratory emissivity data, Science Advances 7 (2021) eaba9428, https://doi.org/10.1126/sciadv.aba9428.

CHAPTER 3

Finding and reading planetary data

Michael Aye
Laboratory for Atmospheric and Space Physics, University of Colorado, Boulder, CO, United States

3.1 Data acquisition in planetary science

3.1.1 Introduction

There are several official archives from national space agencies available for freely accessible planetary data sets. The largest ones are those provided by NASA and the European Space Agency (ESA): NASA's "Planetary Data System" (PDS) and ESA's "Planetary Science Archive" (PSA). In addition, the Japanese space agency JAXA also provides a growing archive — following the same archiving scheme as PDS — called "Data and Transmission System" (DARTS). This currently, among many other missons, offers data for the Hayabusa missions, the lunar orbiter Kaguya, the Venus orbiter Akatsuki, and the cometary explorer mission Sakigake.

3.1.2 Data processing levels

Before we dive into the specific archives, it is helpful to understand the standard levels of processing that planetary data go through once the data has arrived on Earth. An added complication here is that the PDS (and archives that follow the same schema) is currently transitioning from the older data format, PDS3, to the new format, PDS4. Most of the archive's data sets are still in the older PDS3 format, but since approximately 2012 all recent and ongoing missions (e.g., NASA's Maven mission to Mars) have been submitting their data formatted as PDS4. Therefore it is required to be aware of the subtle differences in the structuring of these formats from the processing point of view.

The data processing levels for PDS3 are unfortunately weakly defined, so they can differ between different instruments. Example definitions for the Lunar Orbiter Laser Altimeter (LOLA) are shown in Table 3.1. For the machine learning practitioner, the most interesting data levels may be geolocated and pregridded datasets that enable easier combination with other

Machine Learning for Planetary Science
https://doi.org/10.1016/B978-0-12-818721-0.00010-0
Copyright © 2022 Elsevier Inc.
All rights reserved.

Table 3.1 PDS3 processing levels as defined for the Lunar Orbiter Laser Altimeter (LOLA), on board the Lunar Reconnaissance Orbiter (LRO). Source: Document at PDS Small Bodies.

PDS3 Level name	Description
Packet data	Telemetry data stream as received at the ground station, with science and engineering data embedded.
Level 0	Instrument science data (e.g., raw voltages, counts) at full resolution, time ordered, with duplicates and transmission errors removed.
Level 1A	NASA Level 0 data that have been located in space and may have been transformed (e.g., calibrated, rearranged) in a reversible manner and packaged with needed ancillary and auxiliary data (e.g., radiances with the calibration equations applied).
Level 1B	Irreversibly transformed (e.g., resampled, remapped, calibrated) values of the NASA Level 1A, or possibly Level 0, instrument measurements (e.g., radiances, magnetic field strength).
Level 1C	NASA Level 1A or 1B data that have been resampled and mapped onto uniform space-time grids. The data are calibrated (i.e., radiometrically corrected) and may have additional corrections applied (e.g., terrain correction).
Level 2	Geophysical parameters, generally derived from NASA Level 1 data, and located in space and time commensurate with instrument location, pointing, and sampling.
Level 3	NASA Level 2 geophysical parameters mapped onto uniform space-time grids.

instrument data that, ideally, were mapped onto a grid of the same resolution. In the example of LOLA data this level could either be Level 1C for physical values, mapped onto uniform space-time grids, or, more domain-specific, derived geophysical parameters like derived surface densities on uniform space-time grids, stored as Level 3. In general, data levels higher than 1 are produced from domain-specific inference, requiring an interpretive translation of the physical values into a given geophysical parameter. However, lower data levels may be sufficient as long as instrument-specific idiosyncrasies like detector nonlinearities have been removed (e.g., Level 1A in the LOLA example).

The processing levels of the newer PDS4 format have been simplified somewhat and are therefore being followed more strictly, shown in Table 3.2.

Table 3.2 Data processing levels for PDS4 data. Source: Document at Goddard Spaceflight Center.

PDS4 Level Name	Description
Telemetry	An encoded byte stream used to transfer data from one or more instruments to temporary storage where the raw instrument data will be extracted.
Raw	Original data from an instrument. If compression, reformatting, packetization, or other translation has been applied to facilitate data transmission or storage, those processes will be reversed so that the archived data are in a PDS approved archive format.
Partially Processed	Data that have been processed beyond the raw stage but which have not yet reached calibrated status.
Calibrated	Data converted to physical units, which makes values independent of the instrument.
Derived	Results that have been distilled from one or more calibrated data products (for example, maps, gravity or magnetic fields, or ring particle size distributions). Supplementary data, such as calibration tables or tables of viewing geometry, used to interpret observational data should also be classified as 'derived' data if not easily matched to one of the other three categories.

Summary

When choosing a data set, always make sure to understand which processing level the data set is at.

3.1.3 PDS

NASA's PDS is a federation of several geographically distributed nodes, with six of them being science discipline nodes, focusing on Atmospheres, Geosciences, Cartography and Imaging Sciences (CIS), Planetary Plasma Interactions, Ring-Moon Systems (RMS), and Small Bodies. In Table 3.3 one can see an overview of the science nodes (see https://pds.nasa.gov/home/about/ for a more top view introduction into PDS).

The names of the nodes often can be used to infer a group of observation targets for which data are stored at that specific node, however, the same data might be stored at more than one node due to overlapping science cases. Additionally, some nodes might add a processing level that other nodes do not provide. A good example for this situation is NASA's Cassini

Table 3.3 Abbreviated descriptions of the 6 science data-providing nodes of the PDS. Adapted from https://pds.nasa.gov/home/about/node-descriptions.shtml, see link for full version.

Node Name	Description
Atmospheres Node	The Planetary Atmospheres Node (ATM), all nonimaging atmospheric data from NASA's planetary missions, including higher-order products derived from those data.
Geosciences Node	The Geosciences Node (GEO), data sets that are relevant to the geosciences discipline: the study of the surfaces and interiors of terrestrial planetary bodies. Derived image data, geophysics data, microwave data, spaceborne thermal data and spectroscopy data obtained from NASA's planetary science missions, along with related data from laboratory and field studies.
Cartography and Imaging Sciences Node	The Cartography and Imaging Sciences Node (CIS), also known as the *Imaging Node* or *IMG*; NASA's primary digital image collections from missions to the inner and outer Solar System planetary systems (planets, rings and moons, including icy satellites). Imaging provides the archives, ancillary data, sophisticated data search and retrieval tools, and cartographic and imaging science expertise necessary to develop and fully utilize the vast collection.
Planetary Plasma Interactions Node	The Planetary Plasma Interactions Node (PPI), fields and particle data from NASA's planetary science missions.
Ring-Moon Systems Node	The Planetary Ring-Moon Systems Node (RMS), data sets relevant to outer planetary systems containing planetary rings and/or moons, including the ways the rings and moons interact. Remote sensing data (images, imaging spectrometer data, and occultations) for systems beyond the asteroid belt (that is, Jupiter through Pluto). Services: OPUS, a comprehensive search tool for outer planet systems, various tools for planning and analyzing observations, and coordinating special observing campaigns.
Small Bodies Node	The Small Bodies Node (SBN), planetary science data relevant to research of objects generally described as comets, asteroids, and interplanetary dust including dwarf planets, objects in the Kuiper Belt and the Oort cloud, Centaurs, and small planetary satellites. Included are mission, ground-based, and laboratory data. Subnodes: comet (University of Maryland College Park), asteroid and dust (Planetary Science Institute, Tucson AZ) subnodes. The SBN oversees the Minor Planet Center (Smithsonian Astrophysical Observatory, Cambridge MA), a data center, in the execution of its special tasks mandated by the International Astronomical Union, including the categorization and designation of small bodies.

mission, and specifically the image data of the Imaging Science Subsystem (ISS) camera. The prime designated PDS node for all data of the mission is the Atmospheres node, while the RMS node has most of that data as well. However, only the RMS node has a calibrated version of the ISS data available.[1]

Note

Some data exist on more than one PDS node, but usually only one node is the official archive; however, some nodes might offer a higher processing level!

It is also worth pointing out that the search engines of different nodes might be of different quality or ease of use. As an example, the RMS node has a well designed, easy-to-use search interface called *Outer Planets Unified Search* (OPUS) that offers both a beginner-friendly interface and advanced features as an application programming interface (API) to automate searches and downloads, as well as shareable search settings via the current URL of displayed search results).

Another example is the feature-rich image search of the Planetary Image Atlas at the CIS node at https://pds-imaging.jpl.nasa.gov/search. This search interface also offers a content-based image search for selected instruments; however, these searches are implemented as a beta test only so far.

3.1.3.1 Organizational structure within a node

Within a science discipline node, data storage is often, but not always, first structured via the target of interest, for example Mars or Mercury, and underneath that by name of the planetary mission that acquired that data, for example *Mars Reconnaissance Orbiter* (MRO) or the *Mercury Surface, Space Environment, Geochemistry and Ranging* mission (MESSENGER). See https://pds-geosciences.wustl.edu/missions/mep/index.htm for an example of the Mars exploration page of the Geosciences Node. Let us drill down the structure tree, using MRO as an example. Clicking on its entry[2] in the list provides a link to search tools: the *Mars Orbital Data Explorer (ODE)*, which is older but more complete on all the mission data, and

[1] The original Cassini PDS contract did not require to deliver calibrated data, only the tools required to perform the calibration yourself.

[2] https://pds-geosciences.wustl.edu/missions/mro/default.htm.

Instrument *Follow links to instrument team web sites*	**PDS Archives**
CRISM (Compact Reconnaissance Imaging Spectrometer for Mars)	**CRISM Archive** **CRISM Spectral Library** **CRISM Type Spectra Library**
SHARAD (Shallow Radar)	**SHARAD Archive**
Gravity / Radio Science	**Gravity/RS Archive**
HiRISE (High Resolution Imaging Science Experiment)	**HiRISE Archive** *at the Imaging Node*
CTX (Context Imager)	**CTX Archive** *at the Imaging Node*
MARCI (Mars Color Imager)	**MARCI Archive** *at the Imaging Node*
MCS (Mars Climate Sounder)	**MCS Archive** *at the Atmospheres Node*
Accelerometer	**Accelerometer Archive** *at the Atmospheres Node*
SPICE (Geometry and Navigation)	**SPICE Archive** *at the NAIF Node*

Figure 3.1 Table of available instruments and archives for a mission. Partial screenshot from https://pds-geosciences.wustl.edu/missions/mro/default.htm.

the newer *Planetary Image Atlas* that focuses on image data of the HiRISE, CTX, and MARCI instruments, which is run by the CIS node.

Below the two search engines we find a list of all the instruments on MRO, as shown in the screenshot in Fig. 3.1, where the resolved or explained instrument acronyms in the left column should be sufficient for a new user to distinguish between the different kinds of data: spectral, radar, gravity, or image data. Many machine learning tasks will be using image data, so let's continue with the HiRISE experiment. Clicking on the archive link for the HiRISE experiment takes us to the CIS Node, with all image data of the MRO mission (including HiRISE, CTX, and MARCI) since the CIS node is focusing on archiving image data.

From there we click on Online Data Volumes where we can see two more subcategories of how planetary mission data are being archived: *Releases* and *Volumes*.

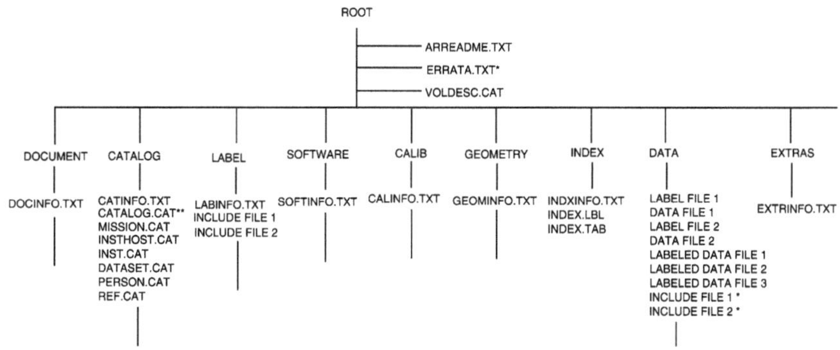

Figure 3.2 PDS3 Volumes layout. From Fig. 19.1 of the PDS3 standard document.

Releases and volumes

Releases and Volumes are the older structural entities of the still much used PDS3 archives. *Releases* are created whenever a data producer delivers a set of data to the PDS. Each *release* consists of one or more PDS3 *volumes*. Volumes refer to the historical definition of a PDS3 unit of storage consisting of a physical medium like a CD, DVD, or magnetic tape, from the times before fast online data transfer when data was still delivered to the PDS via mail; though these media are no longer used, all PDS3 data are still structured in this way.[3] Volumes are always set up with the same directory structure with folders for calibration data, documentation, software, the actual data files, and others (see Fig. 3.2). Which folders are important for the end user may differ for the specific data analysis tasks, but here are some useful files and folders to keep in mind:

AAREADME.TXT Alphabetized for a reason to appear first in the directory listing, this file is a must read to understand the data release structure quickly.

ERRATA.TXT Worth a quick check, because identified problems with a release will be announced in this file.

DOCUMENT This folder may contain the so called *Software Interface Specification (SIS)* documents for the data product in question. This is the reference document provided by the data producer, detailing what each data element in the delivered data depict and describing

[3] Disclaimer: The author describes his understanding of the PDS archives from a user perspective, and definitions may not necessarily be complete or precise from the archivist's point of view.

any caveats that are required to know when working with the data. See the HiRISE SIS as an example: https://hirise.lpl.arizona.edu/pdf/HiRISE_RDR_SIS.pdf (see meaning of *RDR* discussed below).

> **Note**
>
> The Software Interface Specification (SIS) document is the most important document to read for understanding how a data set was prepared and what each data element means. This document is available for all data delivered to and archived at the PDS.

SOFTWARE Sometimes this folder contains tools required to open or analyze the provided data, but very often these programs rely on very old software libraries that are either hard to find or install nowadays. For this reason, the author is Co-Investigator of a new NASA project to create a general PDS data reader for Python (See below "Spaces to watch", Section 3.1.6).

GEOMETRY This folder may contain important observation geometry data required to analyze the data properly. Most (planetary) data is influenced by its circumstances in which the data was taken, (a.k.a. metadata). In the case of observational geometry data, it describes the colocation of the Sun, the observed body, the sensor that recorded the data, and any other body that might influence the data, over time. Today, these data are mostly provided as so called SPICE kernel files. SPICE[4] is a computational framework for solar system geometries used by NASA, ESA, and other space agencies.

INDEX This folder contains text-based index files with some of the available metadata organized in columns and one row per data element of this volume. These index files provide a good way to identify and filter data files of interest (see Section 3.1.6).

DATA This is where the actual data files are stored. Both PDS3 and PDS4 store a lot of important metadata in so called label files that have the same file name but with the extension ".LBL" instead. These files always need to be colocated with the associated data file, because the label file points to it in form of a relative path.

[4] https://naif.jpl.nasa.gov/naif/.

EDR and RDR

Older but still often used terminology is the *Experimental Data Record (EDR)* and the *Reduced Data Record (RDR)* (see for example the archive structure of HiRISE data). In short, EDR data are the uncalibrated raw data records in "digital number" units (i.e., digitized read-outs of the analog sampling devices), while RDR data have been calibrated and reduced further to a varying degree depending on the use case. Details for each product should always be available in the corresponding SIS documents, e.g., the HiRISE EDR SIS and the HiRISE RDR SIS.

PDS4 collections and bundles

PDS4 is not using the older structure of volumes and release and instead uses collections and bundles. A PDS4 collection is a grouping of related data products. For example, a collection could contain all the calibrated data from a given instrument, or the data from a specified mission phase. A PDS4 bundle contains a grouping of related PDS4 collections. In the below link, for example, the OSIRIS-REx OCAMS bundle contains collections for the raw, reduced, and calibrated data, calibration data, and documents, for the OCAMS instrument.

Source: Key terms defined at the Small Body Node.

Note

Both PDS3 and PDS4 file content access is done via the associated label file for each data file (see Section 3.1.5). Always download the associated label files as well and place them in the same folder as the data files.

3.1.4 ESA's Planetary Science Archive

The Planetary Science Archive (PSA) of the European Space Agency (ESA) is currently a catch-all archive for all mission data with relation to ESA. Its start page (https://archives.esac.esa.int/psa/) allows users to filter for a mission, an instrument, or a target body, upon which a table view will be shown with the search filter applied to the overall PSA database. Fig. 3.3 shows an example table view after filtering for the Rosetta mission.

One is now able to filter the data further down with the widgets on the left side, e.g., by instrument, data type (images, spectral, etc.), or processing

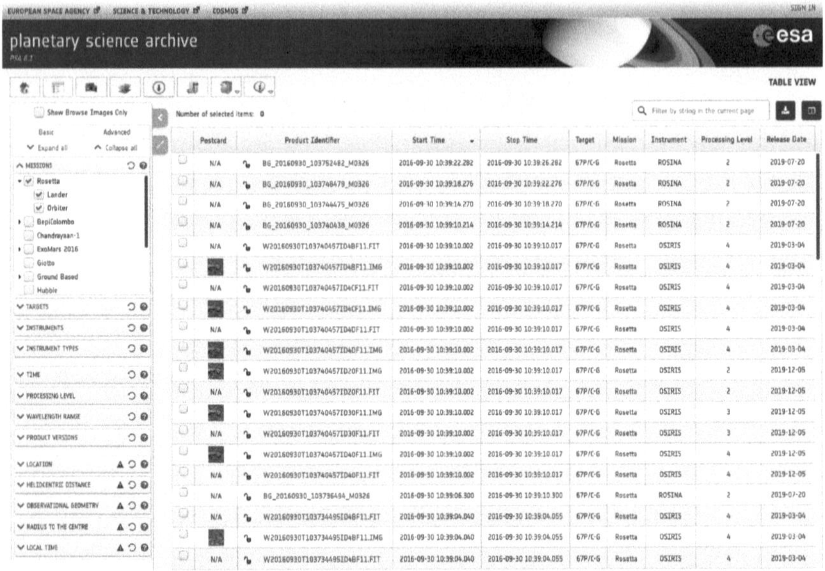

Figure 3.3 Screenshot of first table view at the Planetary Science Archive when initially filtered for the Rosetta mission.

levels. Note that the PSA follows the PDS3/PDS4 processing level distinctions as is evident by the help text shown when clicking on the question mark for the processing level section.

A defining difference compared to PDS is that the PSA is also used for archiving data that has not yet been released to the public. These data will still appear in search results, but with a locked icon in front of the product identifier, indicating that only a login with access permissions would allow the showing of the preview "postcard" (in PSA terms) and the download of these data. Clicking on a product identifier will open a display for related metadata, like data acquisition times, related documentation, and observational geometry, if available.

After selecting several product IDs by placing hooks in the first column of the table, by clicking on the download action button in the upper right next to the search input field they can either be downloaded directly or sent to the *Download Manager*. The user can then search for other data of interest so that everything can be downloaded together at the end of a search session.

3.1.5 Reading data with Python

"Reading data" in Python usually means reading the data into a numpy array that could then be further processed using other powerful analysis libraries like *pandas*[5] or *xarray*.[6] Some data can also be read directly into pandas, for example using `pandas.read_csv` for text input files, or into xarray using `xarray.open_rasterio` for image data, if the rasterio/GDAL reader supports the given image format.

However, the overall situation with reading planetary data is tricky. The PDS3 format allows many different approaches to storing data. This made it very difficult to create a general reader that would be able to read all the different kinds of data stored at the PDS, so no complete solution exists in any programming language. Prototypes of general solutions in Python exist, for example *planetaryimage*,[7] but were abandoned due to lack of funds and/or time required to solve all the problematic cases identified. At this[8] URL, the author of the planetaryimage package shows that more than 50% of all image data stored on the PDS cannot be read using a standard approach. This situation finally led to a new, recently funded project to create a universal PDS reader, see Section 3.1.6.1 for more.

In the mean time, *GDAL*[9] — a widely used translator library for raster (i.e. image) and vector data — also added drivers for PDS3 and PDS4 data, with recent improvements of reading capabilities. Despite GDAL focusing on data that was projected onto a geo-spatial surface, it still works for unprojected data (i.e. lower processing levels) as well, if one is only interested in reading the image pixel data itself.

Note

It is recommended to use the higher level, GDAL-wrapping library *rasterio*[a] instead of using GDAL directly, as it provides a more modern and Pythonic interface.

[a] https://rasterio.readthedocs.io.

[5] https://pandas.pydata.org/docs/.
[6] https://xarray.pydata.org/en/stable/.
[7] https://github.com/planetarypy/planetaryimage.
[8] https://planetaryimage.readthedocs.io/en/latest/supported_planetary_image_types.html.
[9] https://gdal.org.

In the case of PDS4 data, the existing `pds4_tools` package[10] can be used. It should work for all PDS4 data, and the authors are interested in any issues found while trying to read PDS4 data.

3.1.5.1 Example reading of PDS3 data

Download the example Cassini ISS image showing a small part of Saturn's rings from here: https://opus.pds-rings.seti.org/opus/#/view=detail&detail=co-iss-n1880813932. Remember that you need to download both the *.IMG* and *.LBL* file and store them next to each other in the same directory! In the following example, there is a folder where all data for this image ID is stored. Note how one must give rasterio the path to the label file and not directly to the IMG file, because the label file contains information for the driver to read the file properly, and then points eventually to the IMG file for reading.

You can try the following in an IPython or Jupyter notebook session, for example. The >>> characters indicate where you type in the python statement (i.e. the "prompt"), lines without the >>> are output. For a notebook session, you can also copy the whole code-block and paste it into a notebook cell, including the prompt characters that will be automatically ignored.

Listing 3.1:

```
>>> import rasterio
>>> folder = "/home/maye/Dropbox/data/ciss/db/N1880813932/"
# the .IMG file is in the same folder!
>>> fname = folder + "N1880813932_1.LBL"
>>> dataset = rasterio.open(fname)
[...] NotGeoreferencedWarning: Dataset has no geotransform, gcps, or rpcs. The
      identity matrix be returned.
```

At this point, rasterio will emit the "NotGeoReferencedWarning". This is caused by the image being raw data without any projection onto a target body. This is fine as long as one only wants to get to the binary data of the image file for now. To get to the binary data, we use the `read` method of the dataset object:

Listing 3.2:

```
>>> img = dataset.read()
>>> img.shape
```

[10] https://sbnwiki.astro.umd.edu/wiki/Python_PDS4_Tools.

```
(1, 1024, 1024)
```

This results in `img` being a numpy array with the shape of `(1, 1024, 1024)` — and not (1024, 1024) — because raster data usually comes in multiple bands or layers of the same spatial area. A standard color image for example has three bands, for the red, green, and blue channel. To plot this using `matplotlib.imshow`, we need to select the band using index 0, to get a 2D array with the shape of `(1024, 1024)`:

Listing 3.3:

```
>>> import matplotlib.pyplot as plt
>>> plt.imshow(img[0], cmap='gray')
>>> plt.colorbar()
>>> plt.show()
```

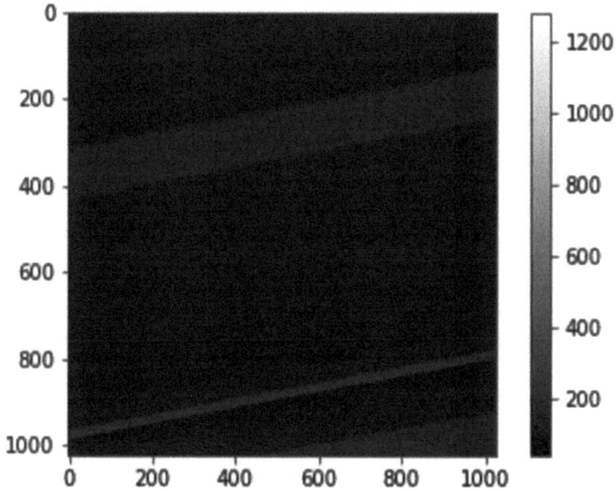

As one can see the plot has extremely bad contrast—what happened? This is mostly caused by extreme "cold" or "hot" pixels that span the color stretch too far to allow seeing more details. Let's constrain the color stretch to the 2nd and 98th percentile of the pixel values, respectively:

Listing 3.4:

```
>>> import numpy as np
>>> p2, p98 = np.percentile(img, (2, 98))
>>> plt.imshow(img[0], vmin=p2, vmax=p98, cmap='gray')
>>> plt.colorbar()
>>> plt.show()
```

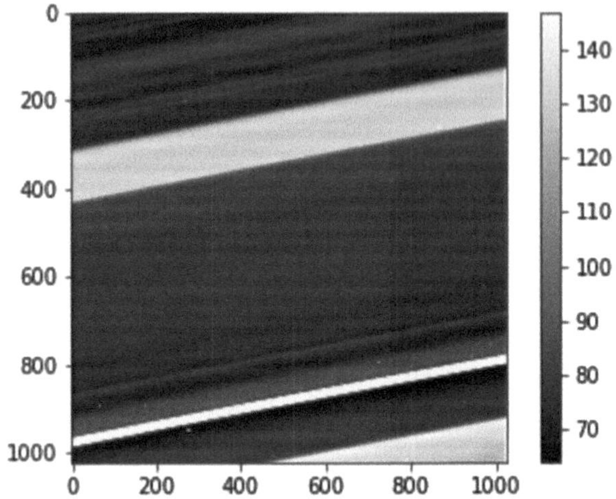

One can see the reduced range of pixel values on the colorbar.

3.1.5.2 Troubleshooting data reading

What can one do if both `rasterio` and `planetaryimage` won't open the data of interest? It is too difficult to provide a general solution, but one strategy that might work is to use the binary numpy reader `np.fromfile()`.[11] This function requires that you provide a numpy datatype, like `float16`, which determines the stepping of number of bytes being read for each data value. Many older PDS3 data were stored in one straightforward binary data blob for simplicity. By inspecting the label file carefully for any hints on specific binary storage information, one often can find out the datatype that would need to be provided to `np.fromfile()`. Failing these efforts, one can always join the OpenPlanetary community and find help there (see Section 3.1.6.3).

3.1.6 Spaces to watch

3.1.6.1 PDR

Recently, NASA funded a new project of a *Planetary Data Reader (PDR)* with the PI Chase Million of Million Concepts and the author of this chapter as one of several Co-Investigators to create a PDS data reader that will seamlessly wrap existing reader utilities for different data into one interface for the end user. Monitor our progress at this GitHub repository: https://github.com/MillionConcepts/pdr.

[11] https://numpy.org/doc/stable/reference/generated/numpy.fromfile.html.

3.1.6.2 PlanetaryPy

The author is part of an initiative to collect and combine planetary science related Python tools into a core *planetarypy* package. One of the useful tools that exist already is to receive above mentioned index files and ingest them into easier to use Pandas dataframes (a popular data analysis library, see https://pandas.pydata.org). Currently, the author's first attempt at this package is in beta testing and documented at https://michaelaye.github.io/nbplanetary/, but will soon be migrated to our community effort space at https://github.com/planetarypy/planetarypy.

3.1.6.3 OpenPlanetary

A large community of data users in planetary science has gathered under the nonprofit organization OpenPlanetary (https://www.openplanetary.org), of which the author of this chapter is a board member and other editors and authors of this book are members. The organization offers a Slack chat for real-time support, with a multitude of channels focusing on diverse topics like Jupyter notebooks, conda, PDS, and PSA, to name just a few. It is free to join the organization and its Slack chat via https://www.openplanetary.org/join. For increased outreach, there is also a public forum where issues can be discussed without having to join Slack, which can be found at https://forum.openplanetary.org.

CHAPTER 4

Introduction to the Python Hyperspectral Analysis Tool (PyHAT)

J.R. Laura[a], L.R. Gaddis[a,c], R.B. Anderson[a], and I.P. Aneece[b]
[a]U.S. Geological Survey, Astrogeology Science Center, Flagstaff, AZ, United States
[b]U.S. Geological Survey, Western Geographic Science Center, Flagstaff, AZ, United States
[c]Universities Space Research Association, Lunar and Planetary Institute, Houston, TX, United States

Key points

- The Python Hyperspectral Analysis Tool (PyHAT) allows users to visualize and analyze planetary spectroscopic data.
- We illustrate use of PyHAT orbital to analyze CRISM, M^3, and SELENE data, and PyHAT in-situ to analyze LIBS data.
- PyHAT is a user-friendly, open-source way to run otherwise costly, complicated, or programming-intensive methods.

4.1 Introduction

Spectroscopic data are rich in information and commonly used in planetary research for purposes such as remote surface mineral mapping, derivation of thematic products for scientific analysis and exploration, and in-situ quantification of mineralogy and chemistry. For analyzing reflectance spectral data, instrument teams often use simple equations to derive thematic products (e.g., iron oxide abundance using Clementine data [28], olivine abundance using band center analysis of Moon Mineralogy Mapper (M^3) [22]). Often these equations are either unavailable to the public or coupled to expensive software tools such as MATLAB®, Interactive Data Language (IDL), or Environment for Visualizing Images (ENVI). For quantification of in-situ data, the methods involved are not always made publicly available, and when they are made available they may be more complex than a scientist outside the instrument team can implement. This manuscript describes the Python Hyperspectral Analysis Tool (PyHAT) Python library and associated graphical user interfaces (GUIs) that enable spectral data visualization, thematic image derivation, and both simple and sophisticated

Machine Learning for Planetary Science
https://doi.org/10.1016/B978-0-12-818721-0.00012-4
Copyright © 2022 Elsevier Inc.
All rights reserved.

spectral processing and analysis methods in a cross-platform, open-source environment.

The development of PyHAT has been motivated by three drivers. First, we note a general lack of freely available and open-source implementations of published spectral analysis equations. For example, while the Mars Reconnaissance Orbiter (MRO) [57] Compact Reconnaissance Imaging Spectrometer (CRISM) [32] team provides both the mathematical formulation of equations used to generate derived products [50] and the CRISM Analysis Toolkit (CAT) ENVI tools, which contain executable implementations of those equations, use of the tool requires a paid ENVI license. Likewise, data from the M^3 sensor have been used to generate derived data products [34], but we were not able to locate a publicly available implementation for use by the community. Second, in our experience, analysis of multi- and hyperspectral data requires an analytical environment that supports exploratory data analysis [48]. The derived products defined by the CRISM and M^3 teams are similar, but differences in the sensors, target body, and science goals result in slightly different implementations. These differences required iteration to determine the best possible bands and band parameters to use for derivation of a given product. Likewise, the in-situ component of PyHAT has been driven by the desire to apply statistical learning algorithms to rover-collected hyperspectral data to both visualize and manipulate large numbers of spectra and to accurately quantify target composition. Third, as a team that works with both orbital and in-situ hyperspectral data, we sought to centralize the necessary bookkeeping and preprocessing algorithms. For example, the underlying data structures and frequently used operations, such as spectral smoothing or continuum removal, are agnostic to the sensor used to capture the signal. The methods do not use information about the instrument or the type of spectrum; they just apply a mathematical operation to the numerical values of the spectrum itself. This assertion is true even for more sophisticated methods such as dimensionality reduction or multivariate regression.

To make PyHAT as accessible as possible, the PyHAT library is written and documented using standard Python development practices, and supports a functional programming paradigm on top of the ubiquitous Pandas Data Frame. Additionally, we maintain NumPy style docstrings on all of our user facing application programming interface (API) so that users and developers can easily incorporate PyHAT into other projects. The PyHAT orbital API is especially powerful within the Jupyter analysis environment [25] to perform exploratory data analysis tasks. For example, a common workflow might include continuum fitting to a number of different spectra, and we find Jupyter notebooks to be an ideal place to iteratively test different fits due to the rapid iteration afforded in the notebook environment. Two GUIs were also designed to support a user friendly approach that assumes no programming knowledge. The first, a Quantum Geographic Information System (QGIS) plugin, supports the generation of several standard, derived data products from orbital mapping spectrometer data. The second GUI is a full featured, stand-alone application to make use of the PyHAT statistical and machine learning capabilities with an emphasis on in-situ data collected by point spectrometers, primarily in the laboratory and on landed spacecraft. The GUIs do not have already-ingested data; rather, they allow users to load their own data.

The remainder of this chapter is organized as follows. In Section 4.2, we describe the PyHAT library architecture. In Section 4.3, we document the PyHAT orbital architecture and briefly describe the GUI that exposes components of the library. We illustrate the use of PyHAT orbital, along with the QGIS and Jupyter notebook interfaces to analyze CRISM data (Section 4.3.1), M^3 data (Section 4.3.2), and Kaguya Spectral Profiler (SP) data (Section 4.3.3). In Section 4.4, we illustrate the use of the PyHAT in-situ GUI for statistical learning and demonstrate both data preprocessing, exploration, and analysis. Finally, we present our intended future directions for the project in Section 4.5.

4.2 PyHAT library architecture

The PyHAT library (imported in Python as "libpyhat") is partitioned into preprocessing and input/output capabilities, general spectral transformation that focus primarily on preprocessing operations, algorithms and analysis tools for working with mapping spectrometer data, algorithms and analysis tools for working with point observations, and finally statistical and machine learning analysis methods. In Fig. 4.1, we graphically illustrate the decomposition of the library using the library module names.

As previously indicated, the data structures used throughout PyHAT for spectra are based upon Pandas Data Frames. A Data Frame is a two-dimensional table of data with labeled rows and columns. The Pandas library provides a robust set of tools for indexing, labeling, managing, and analyzing two- and three-dimensional (time-series) data. Pandas does not have explicit support for spectra data. Therefore, PyHAT extends (subclasses) the Data Frame. PyHAT uses a "multiindexing" method to work

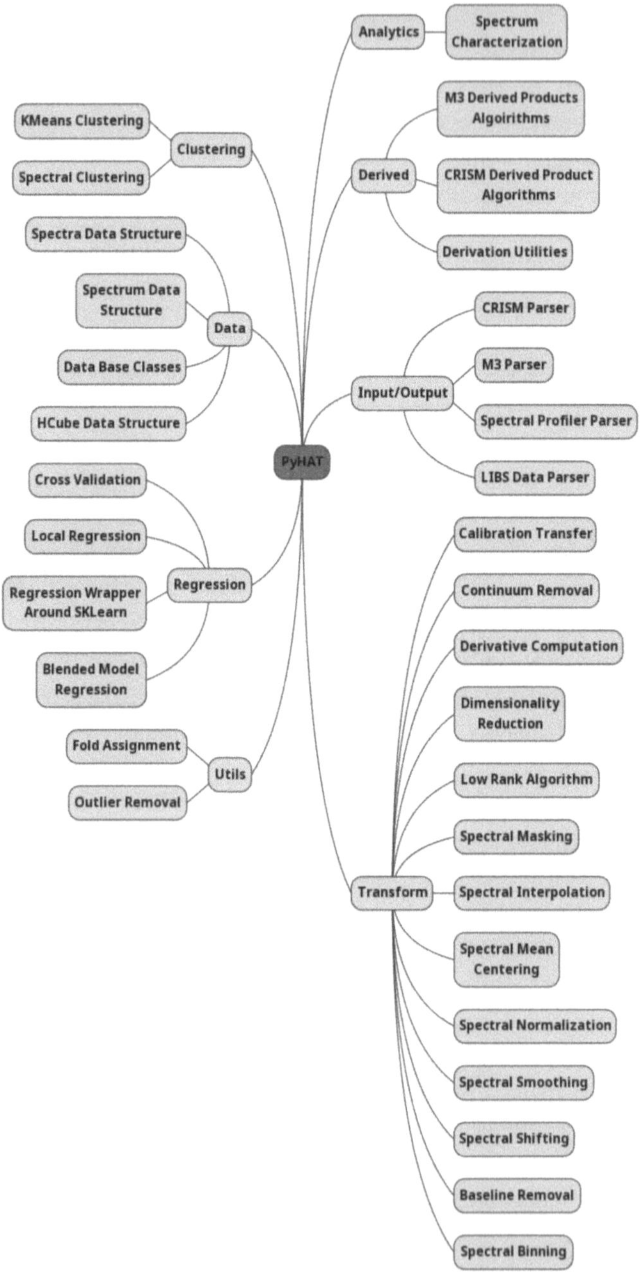

Figure 4.1 PyHAT library architecture composed of eight independent modules. The second level of decomposition highlights capabilities within each module and illustrates the functional decomposition of the library.

with spectral data within Pandas. To store imaging spectrometer data in an intuitive way, PyHAT extends the Pandas Data Frame with support for fuzzy indexing (approximate wavelength) and the addition of convenient analytical tools (described in Section 4.3). The base PyHAT data types (Spectrum, Spectra, and HyperCube) are the foundational data structures upon which PyHAT operates. A HyperCube object sits atop the Geospatial Data Abstraction Library (GDAL) library. In the parlance of Pandas, the Spectrum object is a Series (a single vector of information), the Spectra is a Data Frame (an array of information with the same nominal structure), and the HyperCube is an n-dimensional array of data. From an API consumer perspective, these data structures offer the easiest entry to access the capabilities of PyHAT. From a user perspective, these data structures will be largely transparent.

The library is written such that the data structures are passed to algorithms, which then operate on them. This is a traditional functional programming paradigm. We believe that this supports better function chaining as data are manipulated through some arbitrary number of processing steps: for example, preprocessing to smooth, preprocessing to resample, preprocessing for continuum removal, and finally the application of some analytical method. We have included many of the analysis functions as methods on the base data structures to make interactions from a user perspective even more seamless. We anticipate that some subset of the user group will want to utilize Python's robust introspection capabilities and methods on the data structures that support this use case.

To ensure that the PyHAT library would remain decoupled from any visualization capabilities, which evolve significantly faster than analytical methods, GUIs have been developed for visualization. First, PyHAT was actively developed and used in the now ubiquitous Jupyter [25] notebook environment. This environment is most like a development environment where users have access to the entire library. For derived orbital products, a QGIS plugin has been developed to support geospatial data visualization inside of a geographic information system (GIS) while also leveraging the capability for PyHAT to create high science value derived products. The PyHAT in-situ GUI exposes the statistical learning component of the library for laboratory and landed data. This front-end is a cross platform, PyQT GUI that is installable using simple command-line commands.

We are actively improving the PyHAT library to optimize development of underlying data structures, the implementation of preprocessing algorithms, and the development of what we identify as exploratory analytics

(Fig. 4.1). We have limited the end-to-end pipeline style analysis methods to those published or used by mission teams for the generation of derived products. However, we anticipate growing the analytical methods available within PyHAT to support a myriad of missions and analysis use cases. This is a similar development paradigm to other Python or R analysis packages.

4.3 PyHAT orbital

PyHAT currently explicitly supports two mapping spectrometers and one point spectrometer. Using PyHAT, one can generate standard derived products for two orbital National Aeronautics and Space Administration (NASA) missions: the NASA CRISM from MRO, and the NASA M^3 on the Indian Space Research Organization (ISRO) Chandrayaan-1 (Ch1) mission. These hyperspectral data sets are both delivered as three dimensional image cubes where the x and y dimensions are the spatial components and the z dimension represents the spectral component. The spectral component is labeled by the light wavelength that each band has captured. The generation of derived product is then either the extraction of a single band, as in the case of a single-wavelength radiance product, or the mathematical manipulation of some number of bands, for example a ratio of two bands. The derived products implemented in the PyHAT library have been developed by the respective mission instrument teams to be representative of some surface composition. For example, derived products can be generated to highlight areas of potential iron oxide, carbon dioxide frost, or water ice. The application of a derivation algorithm to a data set has been abstracted in a functional way; a generic algorithm is applied to an arbitrary array of radiance values. Therefore, one can add a new algorithm quite easily. PyHAT also supports the SELenological and ENgineering Explorer (SELENE) Kaguya SP, a hyperspectral point spectrometer.

Below we demonstrate usage of PyHAT using both the Jupyter analysis environment and as a QGIS plugin. PyHAT has been developed on top of the standard Python scientific computing stack [4] and integrates well into the Jupyter environment. This means that scientists used to working with Pandas Data Frames, SciPy statistical analysis capabilities, and MatPlotLib's visualization suite can immediately explore spectral data in a familiar analysis environment. QGIS is an industry-standard, open-source GIS with a wide user base, planetary data and projection support, a robust development community, multithreaded image rendering that support rapid visualization of large data sets, and cross platform support. Importantly, QGIS is also free and open source. QGIS utilizes a plug-in style architecture to facilitate outside contributions and expansion of QGIS functionality. We have developed the QGIS plug-in to support the derivation of derived thematic products for CRISM and M^3. In implementation, the QGIS plug-in is a thin wrapper leveraging the PyHAT API and when possible the QGIS API for file input/output and rendering.

The QGIS interface (Fig. 4.2) for both M^3 and CRISM sets of PyHAT derived parameters is implemented as a simple pull-down menu of operations. This allows the user to select each of the implemented parameters in turn, calculate the derived or parameterized images and add them to an active image list, stretch them, and combine them into Red Green Blue (RGB) color composites that closely resemble those implemented by the CRISM and M^3 science teams. Differences in the results of PyHAT and those of the science teams for any given algorithm typically result from a lack of image smoothing or spatial filtering in PyHAT results, and the use of slightly different standard stretch algorithms in PyHAT. As shown below, the results for CRISM and M^3 parameters match quite well; additional, more detailed comparisons of the output are available on the PyHAT GitHub documentation.[1] As noted by both science teams, the derived parameter images represent a method of reducing spectral variation within a single scene into a smaller, limited number of dimensions, and they should be used as quick-look assessments of surface mineralogy. Actual compositional interpretations of surface mineralogy should be based on more detailed examinations by comparing the spectra in the source CRISM and M^3 data to a relevant spectral library.

4.3.1 Compact Reconnaissance Imaging Spectrometer for Mars (CRISM)

The CRISM instrument on the NASA MRO is a hyperspectral imager that acquires data at Visible (VIS) to Near-InfraRed (NIR) wavelengths (0.37–3.92 μm at 6.55 nm/channel, 544 channels) for Mars [32,33]. Targeted observations by the CRISM instrument are acquired at 18 m/pixel spatial resolution (Full Resolution Targets (FRT)) or 36 m/pixel spatial resolution (Half Resolution Targets (HRT)). The multispectral mapping mode used 72 channels at 200 m/pixel. The spectral range covers electronic transition absorptions due to olivine and pyroxene (dominant phases in Mars' basaltic crust) and their ferric alteration products (e.g., [1,11,43,46]),

[1] https://github.com/USGS-Astrogeology/PyHAT_Point_Spectra_GUI.

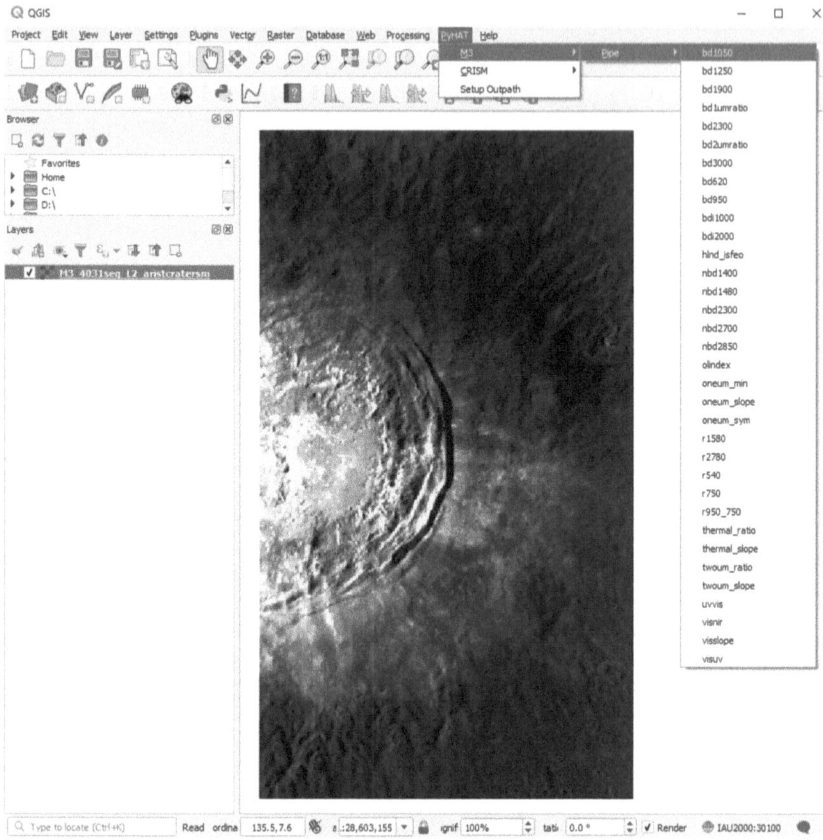

Figure 4.2 QGIS interface showing a M^3 frame of the eastern side of Aristarchus crater floor and highlighting the PyHAT pull-down menu for the M^3 spectral parameters.

and vibrational absorptions in H_2O-, hydroxyl-, and carbonate bearing alteration products including phyllosilicates, sulfates, oxyhydroxides, and carbonates [10,12,19,20,39]. Higher-level CRISM data available in the Planetary Data System (PDS) have been radiometrically corrected (e.g., correction for the solar irradiance, removal of stripes and spikes, see [33]). The CRISM science team published algorithms and derived image products based on multispectral parameters that enhance specific mineralogic signatures [36]. These CRISM summary products were incorporated into the IDL-based CAT [31,42] supporting standard CRISM processing by the scientific community. The CAT is an effective tool for rapid assessment and visualization of an information-rich data set. More recently, a revised set of spectral parameters has been released that includes new parameters, validated by the CRISM science team, designed to capture the diversity of surface mineralogy on Mars [50]. Parameters that highlight the presence of dominant mineral classes [e.g., mafic minerals (olivine, Low-Calcium Pyroxene (LCP)/High-Calcium Pyroxene (HCP)), phyllosilicates (Al-, Fe/Mg-rich), sulfates (mono-, poly-hydrated), carbonates (Mg-, Fe/Ca-rich), iron-oxides (e.g., hematite), ices (H_2O, CO_2)] were used by the CRISM team to create products that represent the surface mineralogy. This paper describes a low-cost, open-source tool that can be updated readily as algorithms change or are added in the future.

In PyHAT, we implemented all 60 of the updated CRISM published algorithms [36,50]. For example, we have included widely used spectral indices from CRISM data like the olivine index (OLINDEX, now OLINDEX3, caused by a positive slope from 1 to 1.5 μm due to olivine and Fe-containing phases). We have also implemented band depth at various wavelengths (e.g., 1.9 μm band depth ($BD1900_2$), caused by H_2O bending and stretching vibrations). Lastly, the convexity of the emissivity band near 2.3 μm caused by hydrated phases due to strong H_2O absorptions at 1.9 and 2.4 μm (SINDEX, SINDEX2) is addressed. Many of the parameters are simple to implement, such as ICER2, a simple band ratio of the CRISM 2350 nm band and the 2600 nm band to highlight CO_2 ice on the Mars surface. Where multiple versions of algorithms exist (e.g., SINDEX and SINDEX2), we make both available. CRISM science team members (F. Morgan and C. Viviano-Beck) assisted with validation to ensure that our products match theirs. By keeping and clearly labeling derivation algorithms, this work supports both reproducibility and a natural evolution of capabilities as our understanding of planetary mineralogy improves.

Using the PyHAT QGIS plugin and the geospatial data manipulation tools inside of QGIS, we have generated a number of comparison images. We present examples of these below both to demonstrate the power of the QGIS plugin and to qualitatively illustrate the alignment between current, closed solutions, and our open-source library. This latter component is of exceptional concern to this team as the efficacy of the implementation is of paramount importance. Notably, the input hyperspectral data for each instrument is different. For CRISM, the Map-Projected Targeted Reduced Data Records (MTRDR) data from the NASA PDS[2] are the input needed for PyHAT; these are available in units of radiometrically and photometrically corrected radiance (I/F). For M^3, the data are also available from PDS

[2] https://pds-geosciences.wustl.edu/missions/mro/crism.htm.

in units of reflectance,[3] but they are delivered in long strips of data that typically cover many latitudinal degrees on the lunar surface. For optimal use of these data for a given science application, scientists typically extract a geospatial scene from the strip; this is also recommended before using PyHAT to allow for more efficient processing speeds.

In Fig. 4.3 we show three derived parameter images for a single CRISM MTRDR scene[4] at Jezero crater on Mars. The mineralogy in this scene has been well described by Horgan et al. [17]. For each scene, we show the result from the science team software (the CAT tools[5]) on the left and the PyHAT QGIS result on the right. The InfraRed Albedo (IRA) parameter shows pixel brightness (I/F) at 1330 nm wavelength (R=R1300; G=R1300, B=R1300), which is useful for highlighting surface morphology and location and properties of features and units. Although the contrast is not as strong in the PyHAT image results, both the CAT and PyHAT versions highlight comparable surface morphology; in this case, both show a relatively smooth crater floor (lower right) with a rougher, lobate fan-shaped feature (at left) and lighter toned rim (center, right) and tongue (center) unit to the south and east of rougher units at top. The next parameter, a FALse-color infrared (FAL) rendition (R=R2529, G=R1506, B=R1080), is an enhanced infrared false color representation that highlights differences between the key mineral groups at Jezero crater. Red-orange colors are usually characteristic of olivine-rich material, blue/green colors often indicate clay, green colors may indicate carbonate, and gray/brown colors often indicate basaltic material. Both the CAT and PyHAT parameter results for FAL feature prominent reddish (olivine-rich) central tongue and rim units, with a grayish (mafic or iron-rich basaltic) floor. The third parameter shown is MAFic mineralogy (MAF) (R=OLINDEX3, G=LCPINDEX2, B=HCPINDEX2), in which iron-bearing rocks are strongly highlighted. Olivine and Fe-phyllosilicates appear red (due to a 1.0–1.7 μm bowl-shaped absorption) in the MAF product, while LCP and HCP (with additional ~2.0 μm absorptions) appear green/cyan and blue/magenta, respectively. For full formulations of the CRISM parameters see Viviano-Beck et al. [50]. Both the CAT and PyHAT parameter results for MAF show a bright red (iron-rich olivine and carbonates) central tongue and rim units, with a blue-green (mafic or iron-rich basaltic)

[3] https://pds-imaging.jpl.nasa.gov/portal/chandrayaan-1_mission.html.

[4] Image identifier frt000047a3_07_if166j_mtr3.img, FRT frame at ~18 m/pixel.

[5] Available at https://pds-geosciences.wustl.edu/missions/mro/crism.htm, an IDL plug-in that provides a GUI interface and a list of available parameters.

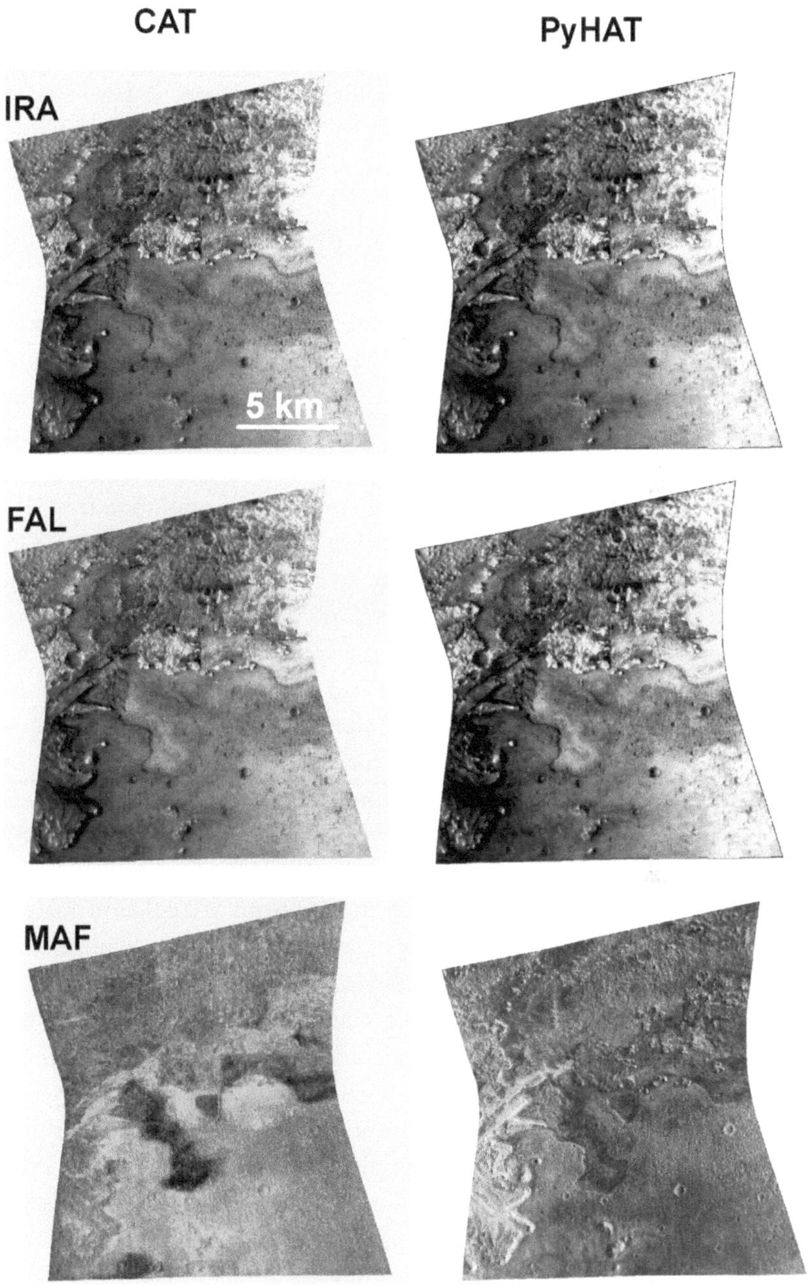

Figure 4.3 CAT (left) and PyHAT (right) results for three derived parameters for a single CRISM MTRDR scene (frt000047a3_07_if166j_mtr3.img, FRT frame at 18 m/pixel) at Jezero crater on Mars: IRA (top), FAL (middle), and MAF (bottom).

floor and a greenish (LCP rich) fan at left. Despite variations in stretches and contrast, the derived parameter images from PyHAT effectively serve as quick-look assessments of surface mineralogy in this scene at Jezero crater on Mars.

4.3.2 Moon Mineralogy Mapper (M^3)

The M^3 instrument flew on the ISRO Ch1 spacecraft (e.g., [6,14,15,38]). The M^3 instrument was a pushbroom imaging spectrometer with a 24° Field of View (FoV) and 0.7 milliradian spatial sampling. Data to be supported include Global Mode (GM) level 1B data from Optical Periods (OP) 1 and 2 and Target Mode (TM) data. GM data from M^3 result from on-board processing of full-resolution TM data to reduced spatial and spectral resolution. TM data have 446 to 3000 nm spectral coverage, 259 channels, 10 nm/band spectral resolution, and 70 m/pixel spatial resolution from 100 km altitude (later reduced to 140 m/pixel at 200 km orbital altitude). GM data have 85 channels covering the spectral range from 460 to 2976 nm, and spatial resolution of 140 m/pixel from 100 km altitude. GM spatial resolution decreased to 280 m/pixel when the spacecraft climbed to a 200 km orbit beginning May 13, 2009 until the end of the mission on August 16, 2009. GM data cover more than 95% of the Moon.

M^3 data are uniquely valuable for lunar surface characterization, for example, surficial water (e.g., [30,38]) and soil and rock mineralogy were mapped at high spatial resolution (140 m/pixel) at wavelengths out to ~2.4 μm (e.g., [5,9,22,24,26,34,35,37,44,47,53]). A series of mineral indicator parameters were developed and used by the M^3 team as a guide to analyses and to provide products that depict lunar surface mineralogy. The parameters were developed to capture the dominant modes of spectral variance related to the presence or absence of mafic silicates, soil maturity, and space weathering. M^3 spectral coverage addresses characterization of the two most common mafic minerals on the lunar surface, olivine and pyroxene. Both minerals have diagnostic absorptions that vary with composition. Pyroxenes have two diagnostic absorptions near 1 and 2 μm (e.g., [1,11]), and olivines have an absorption feature near 1 μm comprised of three separate overlapping mineral absorptions (e.g., [7,45]). Although the M^3 science team created and used derivation mineralogy mapping algorithms similar to those of the CRISM team, they did not publish a comprehensive list. The work described here will address the lack of a list by creating a library of validated derivation tools and making it publicly available.

Like the CRISM example above, we present a series of derived M^3 products in which we compare results from the IDL-based M^3 Tools software (used by the science team but previously not made public) to those from PyHAT. These have also been generated using the user-friendly QGIS interface, using as input an approximately 500 by 300 pixel M^3 subframe[6] that features the eastern portion of Aristarchus crater on the Moon. The mineralogy of this area is described by Mustard et al. [34]. The north rim of the Aristarchus crater has a moderate 1-μm and a stronger 2-μm band depth (attributed to the presence of LCP, typical of a dominantly noritic or upper crustal lunar composition). It has a prominent olivine-rich impact melt deposit (with a strong absorption band at 1-μm and no 2-μm band) extending over the rim to the southeast. Additionally, it has a spectrally featureless impact melt deposit (a likely combination of melted local rock or glass) covering the southern floor, eastern rim, and extending eastward. In Fig. 4.4 we show several black-and-white renditions of the M^3 Aristarchus scene. BD950 or "Band Depth 950", (top part of Fig. 4.4 and Eq. (4.1) from M^3 Tools) is a measure of the 1-μm band depth, and the parameter images both feature a spectrally featureless (dark) impact melt deposit on the floor and southern and eastern crater rims and a pyroxene-rich unit on the crater rim to the north and southeast. The middle image of Fig. 4.4 features OLINDEX (Eq. (4.2) from M^3 Tools), the olivine mineral index (a measure of band depth near 1-μm) and like BD950 it highlights similar units, this time with the olivine-rich (bright) portion of impact melt especially highlighted. The bottom image in Fig. 4.4 shows BDI2000 or the "integrated band depth at 2-μm (or 2000 nm)" (Eq. (4.3) after Mustard et al. [34] and M^3 Tools) and highlights the northern crater rim and the stronger 2-μm band depths there due to LCP. In all equations, we use the shorthand R to indicate reflectance at a given wavelength. Stand-alone numbers are wavelength values in nm. In Eq. (4.3), Rc is a continuum reflectance defined as a straight line across the 2-μm (or 2000 nm) absorption band, 1658 is the first wavelength in a series to be integrated over, 40 is the wavelength interval in nm, and n is the number of channels to be integrated over.

$$BD950 = 1 - \frac{R949}{\left(\frac{R1579 - R749}{1579 - 749}\right) \cdot (949 - 749) + R749} \tag{4.1}$$

[6] Extracted from Image M3G20090209T054031_V01_RFL.IMG.

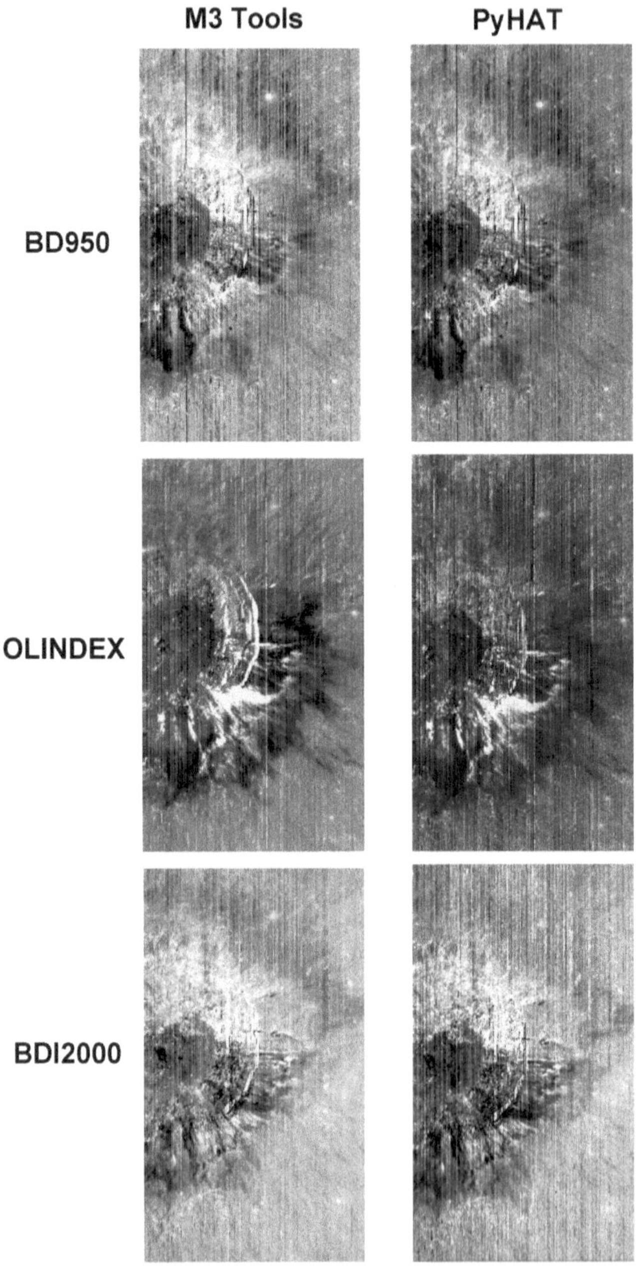

Figure 4.4 M^3 Tool (left) and PyHAT (right) results for three derived parameters from a M^3 subframe 4031 covering Aristarchus crater (40 km across): BD950 (top) or 1-μm band depth, OLINDEX (middle) or the olivine band depth index, and BDI2000 (bottom) or integrated band depth at 2 μm.

$$
\begin{aligned}
OLINDEX = {} & 0.1\frac{[\frac{R1750-R650}{1750-650}](860-650)+R650}{R860} \\
& +0.5\frac{[\frac{R1750-R650}{1750-650}](1047-650)+R650}{R1047} \\
& +0.25\frac{[\frac{R1750-R650}{1750-650}](1230-650)+R650}{R1230}
\end{aligned} \tag{4.2}
$$

$$
BDI2000 = \sum_{n=0}^{21} 1 - \frac{R(1658+40n)}{Rc(1658+40n)} \tag{4.3}
$$

4.3.3 Kaguya Spectral Profiler

The SELENE ("Kaguya") spacecraft, launched in September 2007 and impacted on the lunar surface in June 2009, carried 15 different scientific instruments [41,55]. One of these, the SP [29,55,56], is a hyperspectral point spectrometer comprised of three linear array detectors (VIS, NIR-1, and NIR-12) and two gratings [55]. Detector dependent spectral resolution is 6 nm for the VIS detector and 8 nm for both NIR detectors [16]. The SP data capture 296 spectral bands, with the VIS detector capturing 84 of those between 512.6 nm and 1010.7 nm, NIR-1 capturing 100 bands from 88.3 nm to 1676.0 nm, and NIR-2 capturing 112 bands between 1702.1 nm and 2587.9 nm [16,55]. Spatial resolution varies as a function of orbit height (between 130 km and the nominal 100 km). Given the 125 mm focal length and 0.23° FoV, nominal point resolution is approximately 500 m (along track) by 400 m (cross track) [16].

PyHAT supports reading both the level 2B2 and 2C SP data sets. The level 2B2 data have been radiometrically calibrated and converted to diffuse spectral reflectance. Additionally, the level 2B2 products are also provided with a Kaguya Terrain Camera context image. The level 2C products have had additional spatial correlation, photometric correction, and reflectance conversion algorithms applied.

Below we demonstrate some of the point spectra analysis methods available in PyHAT. In Fig. 4.5 the *libpyhat* library is imported into an active Jupyter notebook session. A few sample files are included alongside the code base both for testing purposes and to support users who wish to experiment with the library before downloading additional data in their regions of interest. The *get_path* function is a helper tool that knows where the data that we provide with the library are located. The final code cell then demonstrates opening the *spc* file that is included in a Japan Aerospace Exploration Agency (JAXA) delivered SP 2C data file. The output *(data)* set is a PyHAT Spectra object that behaves as one would expect a Pandas data frame to behave.

Viewing Spectral Profiler Data

In [1]:

```
from libpyhat.examples import get_path
from libpyhat.io import io_spectral_profiler as sp
```

In [2]:

```
example_path = get_path('SP_2C_02_02358_S138_E3586.spc')
data = sp.open(example_path)
```

In [3]:

```
data
```

Out[3]:

id	0					1					...	36					37				
minor	RAW	REF1	REF2	QA	RAD	RAW	REF1	REF2	QA	RAD	...	RAW	REF1	REF2	QA	RAD	RAW	REF1	REF2	QA	RAD
512.6	5123	0.0402	0.0397	288	27.94	5164	0.0414	0.0409	288	28.77	...	5120	0.0402	0.0398	288	27.72	5046	0.0387	0.0383	288	26.66
518.4	5887	0.0487	0.0482	288	30.99	5955	0.0502	0.0497	288	31.96	...	5881	0.0488	0.0483	288	30.78	5774	0.0468	0.0464	288	29.54
524.7	6375	0.0497	0.0492	288	33.66	6461	0.0513	0.0507	288	34.72	...	6371	0.0498	0.0493	288	33.42	6233	0.0477	0.0473	288	32.06
530.4	6806	0.052	0.0515	288	35.97	6905	0.0537	0.0531	288	37.09	...	6799	0.0521	0.0516	288	35.7	6641	0.05	0.0495	288	34.24
536.5	7494	0.0532	0.0527	288	36.76	7614	0.0549	0.0544	288	37.93	...	7486	0.0533	0.0527	288	36.46	7289	0.0511	0.0506	288	34.96
...	...	...	...	...	...	...	...	...	...	...	...	...	...	...	...	...	...	...	...	...	...
NORMAL_SP_POINT_NUM	38	38	38	38	38	38	38	38	38	38	...	38	38	38	38	38	38	38	38	38	38
UPPER_MARGIN_POINT_NUM	N/A	N/A	N/A	N/A	N/A	N/A	N/A	N/A	N/A	N/A	...	N/A	N/A	N/A	N/A	N/A	N/A	N/A	N/A	N/A	N/A
LOWER_MARGIN_POINT_NUM	N/A	N/A	N/A	N/A	N/A	N/A	N/A	N/A	N/A	N/A	...	N/A	N/A	N/A	N/A	N/A	N/A	N/A	N/A	N/A	N/A
CAL_LAMP_INFO	N/A	N/A	N/A	N/A	N/A	N/A	N/A	N/A	N/A	N/A	...	N/A	N/A	N/A	N/A	N/A	N/A	N/A	N/A	N/A	N/A
MATCHING_ACCURACY_INFO	1001	1001	1001	1001	1001	1001	1001	1001	1001	1001	...	1001	1001	1001	1001	1001	1001	1001	1001	1001	1001

408 rows × 190 columns

Figure 4.5 Jupyter notebook demonstrating loading a SP level 2c file. Here we show the multiindex supporting both per-observations indexing (0, 1, 2,...n) and individual data product indexing (RAW, REF1, etc.). Additionally, the associated, per-observation metadata is presented allowing for operations such as attribute based querying to be possible.

The data being presented in Fig. 4.5 include both spectral information and associated metadata. The axis indices are hierarchical with the x-axis indices indicating the observation identifier at the highest level and then the spectrum type at the minor level. Each SP spectrum is provided with the raw instrument observed values, *REF1* and *REF2* reflectance spectra that have been corrected for mare and highlands respectively, a *QA* or quality control array indicating whether instrument noise is present, and finally a radiance or *RAD* spectrum. The y-axis (wavelengths), while partially obscured to save on overall figure length, illustrates both the observation wavelength and the per-spectrum metadata.

In Figs. 4.6 and 4.7 we demonstrate grouping data associated with individual SP observations and then interrogating the data and metadata associated with a given observation. We have chosen to maintain per-observation metadata even though a subset of the metadata is shared across all observations. We have done this because the overall data volumes are low and this simplifies the bookkeeping necessary to differentiate between metadata at different aggregation scales.

Viewing the data at the observation level

It is frequently preferable to view data at the observation level. To do this we can group the data and then iterate over the observations.

```
In [17]: observations = data.groupby('id', axis='columns')
```

Figure 4.6 Jupyter notebook demonstrating observation grouping so per-point analyses.

One benefit of per-observation metadata is the ability to use Pandas Structured Query Language (SQL) like queries to search a large data frame of Spectra for observations that meet some given criteria. For example, the metadata include latitude, longitude, incidence angle, and emission angle information. Using these metadata, one can query a set of observations (stored in a Spectra object) for only the subset of observations that meet the given criteria.

In Fig. 4.8 we demonstrate the indexing capabilities available on PyHAT Spectrum objects. The columns of the visualized Spectrum object (x-axis) are labeled by wavelength. On data read, some of the wavelength labels can have many trailing zeros (or 0.00000000000004) associated with floating point inaccuracies. These inaccuracies can make label based access (e.g., clip to only visualize between two wavelengths) challenging. For that reason, the Spectra and Spectrum objects support the idea of tolerance. The user can supply a wavelength value within the tolerance and we round within the programming. This makes selection of a subset of the spectra intuitive.

Finally, in Figs. 4.9 and 4.10 we demonstrate a minimal analysis where a linear continuum is fit between 704 nm and 1595 nm and then the band minimum [18] has been computed between 800 nm and 1000 nm.

We hope to have demonstrated the flexibility of PyHAT within the Jupyter notebook visualization. The library supports many more preprocessing and analysis capabilities that we continue to document using Jupyter notebooks on our GitHub page. We anticipate minor changes to the API to improve the user experience, improve interactions between Pandas and our custom Spectra and Spectrum objects, and add extensions to support additional data products.

Viewing the metadata associated with an observation

Once we have an iterator of observations, we can select individual spots to interrogate. Below, we first extract the metadata associated with the first observation in our example 2C data. Below that, we extract the data that we will use for a mock analysis.

```
In [34]:  1 obs0 = observations.get_group(0)
```

```
In [35]:  1 obs0.metadata
```

Out[35]:

id	0				
minor	RAW	REF1	REF2	QA	RAD
SPACECRAFT_CLOCK_COUNT	8.92633e+08	8.92633e+08	8.92633e+08	8.92633e+08	8.92633e+08
VIS_FOCAL_PLANE_TEMPERATURE	21.06	21.06	21.06	21.06	21.06
NIR1_FOCAL_PLANE_TEMPERATURE	18.33	18.33	18.33	18.33	18.33
NIR2_FOCAL_PLANE_TEMPERATURE	243	243	243	243	243
SPECTROMETER_TEMPERATURE_1	18.59	18.59	18.59	18.59	18.59
...	...	...	...	...	...
NORMAL_SP_POINT_NUM	38	38	38	38	38
UPPER_MARGIN_POINT_NUM	N/A	N/A	N/A	N/A	N/A
LOWER_MARGIN_POINT_NUM	N/A	N/A	N/A	N/A	N/A
CAL_LAMP_INFO	N/A	N/A	N/A	N/A	N/A
MATCHING_ACCURACY_INFO	1001	1001	1001	1001	1001

139 rows × 5 columns

```
In [38]:  1 obs0.data
```

Out[38]:

id	0				
minor	RAW	REF1	REF2	QA	RAD
512.6	5123.0	0.0402	0.0397	288.0	27.94
518.4	5887.0	0.0487	0.0482	288.0	30.99
524.7	6375.0	0.0497	0.0492	288.0	33.66
530.4	6806.0	0.0520	0.0515	288.0	35.97
536.5	7494.0	0.0532	0.0527	288.0	36.76
...	...	...	...	...	...
2556.0	6806.0	0.0000	0.0000	296.0	0.00
2564.0	11534.0	0.0000	0.0000	296.0	0.00
2572.0	7317.0	0.0000	0.0000	296.0	0.00
2579.9	11412.0	0.0213	0.0212	288.0	0.34
2587.9	6100.0	0.0000	0.0000	296.0	0.00

269 rows × 5 columns

Figure 4.7 Jupyter notebook demonstrating data and metadata availability within PyHAT for SP data. At the top of the figure we show the metadata attribute and the metadata associated with each data entry in observation 0. At the bottom we show the data attribute which includes just the wavelength and observed DN.

4.4 PyHAT in-situ

For PyHAT in-situ data collected in the laboratory or by instruments on landed missions, PyHAT includes a variety of preprocessing and analy-

```
In [8]:   1 obs0[0, 'REF1'].data[700:1600].plot()
Out[8]: <matplotlib.axes._subplots.AxesSubplot at 0x2b2fc4846e90>
```

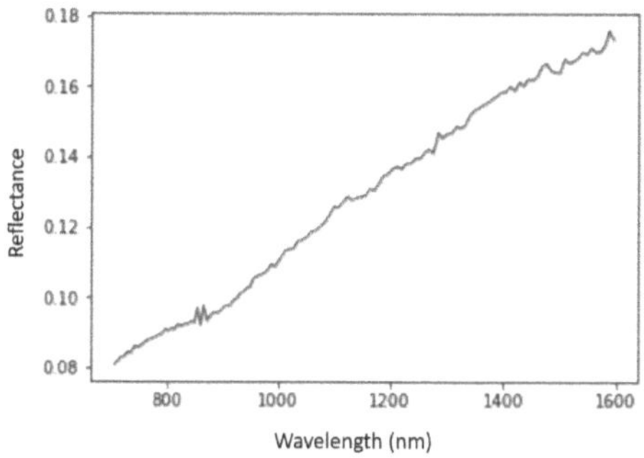

Figure 4.8 Plot of a single observation between wavelengths approximately 700 and 1600 nm inside of the Jupyter notebook environment.

```
In [19]:   1 from libpyhat.transform import continuum as cc

In [29]:   1 corrected, fit = obs0[0, 'REF1'].continuum_correct(nodes=[704, 1595],
           2                                                  func=cc.linear)

In [13]:   1 corrected.data[700:1600].plot()
Out[13]: <matplotlib.axes._subplots.AxesSubplot at 0x2b2fc48e2510>
```

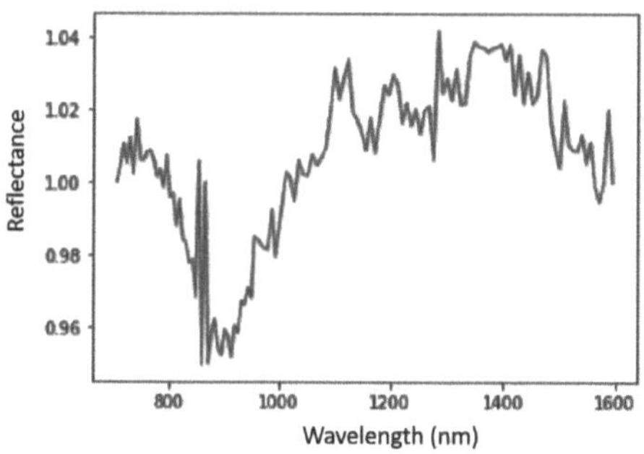

Figure 4.9 An example in which a linear continuum is fit between 704 nm and 1595 nm.

sis methods, accessible either directly through the PyHAT repository, or through a GUI. The source code for the GUI is in a separate repository.[7]

PyHAT in-situ functionality works with data in a simple Comma Separated Value (CSV) format (Fig. 4.11) with each spectrum and its associated metadata stored together in a row. The CSV file has two-level column la-

[7] https://github.com/USGS-Astrogeology/PyHAT_Point_Spectra_GUI.

```
In [34]:  from libpyhat.analytics.analytics import band_minima

In [38]:  minidx, minval = band_minima(corrected,
                                    low_endmember=800,
                                    high_endmember=1000)
          print(f'The band minimum between the ~800 and ~1000nm end members is {minval}.')

          The band minimum between the ~800 and ~1000nm end members is 0.9497539201098775.
```

Figure 4.10 An example of computing the band minimum between 800 nm and 1000 nm.

bels with the top row indicating broad categories of data (e.g., 'wvl' for wavelength, 'meta' for metadata, 'comp' for composition) and the second row indicating more specific categories such as the wavelength of a given spectral channel, the name of the target, or the concentration of a given chemical element in the target. The Pandas library is used to efficiently read data in this CSV format into a data frame, making use of the "multiindex" capability to handle the two levels of column names. Using multiindexed columns enables the software to access large blocks of data (e.g., all of the spectra in the data frame) or individual columns, as needed.

meta	meta	meta	comp	comp	wvl	wvl	wvl	wvl
Target Name	Target Type	Spectrum ID	SiO2	Al2O3	224.666	224.721	224.776	224.831
LANTX1	Llanite	JSC1365	76.9	11.46	35.33	7.33	31.33	-1.67
DIHUQ1	Diopside	JSC1366	54.54	0.7	37.44	20.44	-30.56	20.44
HLNMT1	Pyroxene	JSC1370	41.79	12.32	-10	-7	8	-2
OLIC1	Olivine	JSC1371	50	0.82	5.11	8.11	-12.89	13.11
NANLB1	Labradorite	JSC1372	54.36	27.26	30.56	-10.44	-23.44	5.56
CA9LJ1	Basalt	JSC1373	50.75	13.98	-20.22	-12.22	29.78	12.78
CA9OB1	Obsidian	JSC1374	74.18	13.48	1.44	9.44	2.44	-9.56
CA9KRY1	Rhyolite	JSC1375	67.5	14.44	-6.56	-20.56	-5.56	26.44
MGSDL1	Sodalite	JSC1376	36.96	30.49	-0.61	19.39	12.39	6.39
CA9OB2	Obsidian	JSC1377	69.99	13.27	-21.28	-23.28	-23.28	-14.28

Figure 4.11 Example of the tabular data format used for analysis of in-situ data.

The PyHAT in-situ GUI is organized around the concept of a workflow: a series of discrete data processing and analysis steps referred to as modules (Fig. 4.12). The GUI's main window consists of a central workspace to which these modules can be added using the menu items at the top of the window (e.g., Workflow, Data, Preprocessing, Classification, Regression, Visualization). Modules can be added in any order, giving the user as much flexibility as possible in analyzing their data. The buttons at the bottom of the window (e.g., Re-run, Stop, Insert After, Delete, and OK) allow the user to modify and run the workflow. Above these buttons is a console window displaying module outputs and any errors, and a progress bar.

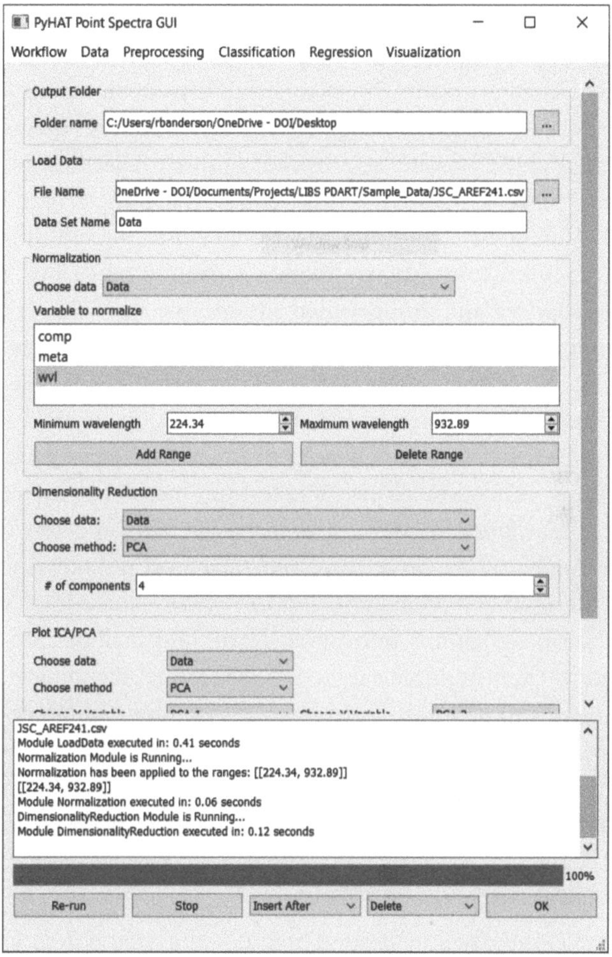

Figure 4.12 Example workflow in the PyHAT in-situ GUI, with modules to specify output directory, load and normalize spectra, run PCA, and plot the results.

PyHAT in-situ includes a variety of tools to help the user with data management, preprocessing, classification, regression, and visualization. Data management tools help with organization of the spectra but do not change the numerical values stored in each spectral channel. Data management tools include reading and writing data, looking up metadata, combining and dividing data sets, removing outliers, and defining folds for later cross validation and testing. The Preprocessing tools, such as spectral masking, baseline removal, calibration transfer, normalization, and dimensionality reduction, change the actual values of data that have been loaded so that they are less noisy, easier to analyze, and produce more accurate results. These changes occur "in-place," modifying the spectra currently stored in memory (but not altering the original source files unless explicitly overwritten). PyHAT in-situ also includes limited classification capabilities, allowing users to separate a data set into different classes (e.g., type of mineral). Currently, K-means and Spectral clustering are implemented, leveraging the scikit-learn library. Work is currently funded to implement additional supervised and unsupervised algorithms.

A major emphasis of PyHAT in-situ is the ability to run regression analyses, allowing users to predict a continuous numerical quantity (e.g., chemical composition) based on spectral data. Regression models are trained on observations for which the quantity of interest has been independently measured. Cross validation allows users to tune model parameters without overfitting. The trained model can then be applied on a test data set to assess the accuracy of the prediction, and finally on unknown data sets to determine the quantity of interest. PyHAT in-situ includes a wide variety of regression algorithms from scikit-learn, and includes the capability to blend results from multiple submodels. Submodel blending is especially useful when the unknown data may span a wide range of target types. For example, blended submodels are used to predict chemical compositions of Mars targets based on spectra from the ChemCam Laser-Induced Breakdown Spectroscopy (LIBS) instrument on the Curiosity rover.

4.4.1 Baseline removal example

In this example, the baseline removal preprocessing step is demonstrated on LIBS data, using several different algorithms available in the PyHAT in-situ GUI. The LIBS technique works by using a high-intensity pulse of laser light to ablate the target, creating a small plasma spark. Light emitted by the plasma is collected and separated into a spectrum. Some LIBS instruments, including ChemCam and SuperCam, have long exposure times relative to the life of the spark and therefore contain significant bremsstrahlung background signal. This background signal contains little useful information, so removing it allows analysis to focus on more informative spectral features of interest [13].

Polyfit is a modification of the most commonly used baseline correction method, polynomial fitting [27]. It finds the baseline by iteratively fitting a polynomial to the spectrum, setting values higher than the polynomial to

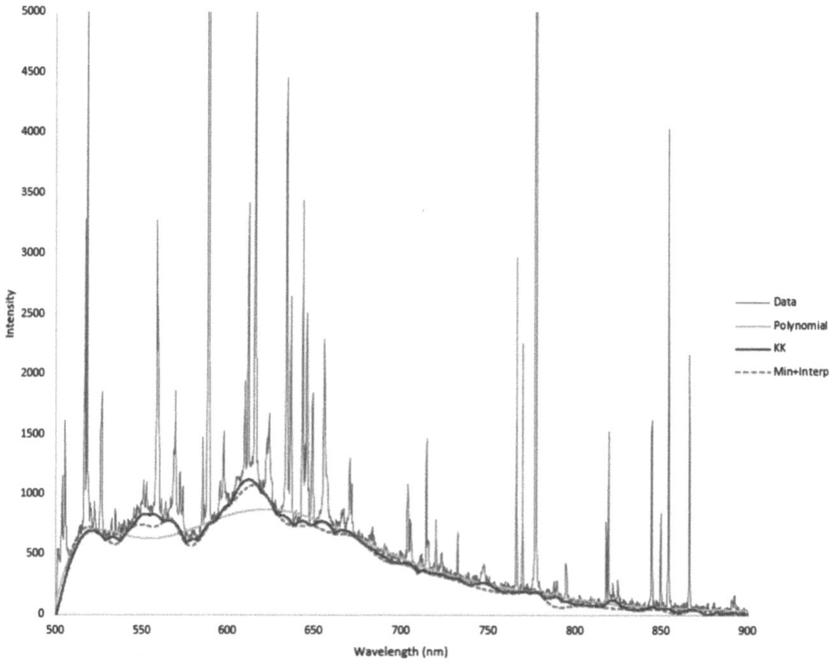

Figure 4.13 Example of baseline removal using polynomial, KK, and min+interpolation algorithms. These are just some of the baseline removal algorithms available in PyHAT in-situ.

the fitted function, and repeating until a threshold of minimum change is reached.

Kajfosz-Kwiatek (KK) fits several polynomials to approximate a signal and then uses the maximum polynomial values to define the baseline. This algorithm was originally designed for x-ray near-edge absorption spectra but is applicable for any data set with many narrow features [23].

The min+interpolation method is a simplified alternative inspired by the Wavelet a Trous + Spline method used by ChemCam. The primary difference is that instead of using a wavelet decomposition to guide the identification of local minima in the original spectrum, this method simply divides the spectrum up into segments of a specified size, finds the minimum in each segment, and interpolates between them. In this example, the interpolation uses a cubic spline.

The outputs from these algorithms are illustrated in Fig. 4.13. In this case, Polyfit misses some of the detail from the original data. The KK algorithm does a relatively good job but has cusps where polynomials are stitched together. The min+interpolate algorithm also performs well in most portions of the spectrum, but does deviate from the expected baseline in some locations. These results use the default settings for each of the algorithms; tuning the parameters would yield improved baseline removal results.

4.4.2 Regression analysis example

After preprocessing the data (e.g., masking bad data, removing outliers, baseline removal, normalization), the user can run classification and regression analyses in the GUI. The goal of regression is to be able to predict a numerical quantity (e.g., composition) based on a spectrum. To accomplish this goal, a regression model identifies correlations between spectral channels and the variable of interest on the basis of training data for which the quantity of interest has been independently measured.

It is possible to train a regression model that predicts the training data with perfect accuracy, but such a model is often too specialized: when applied to novel (unseen) data an overfitted model will give inaccurate results. One well-known example of overfitting is using a high-order polynomial to fit noisy data. To avoid overfitting, the model parameters must be tuned so that the model can handle novel data. This parameter tuning is accomplished using cross validation.

Cross validation involves iteratively withholding subsets of the training data and treating them as unknown data to evaluate the model performance as the parameters are tuned. The stratified folds module, written by coauthor Anderson, is used to define the subsets of data that will be iteratively held out. It works by sorting the training data on the variable of interest and then stepping through the unique values of that variable and assigning spectra to each fold in turn. The result is a set of folds for which the distribution of the variable of interest is as similar as possible. One of the folds is generally set aside as a Test Set: data that are completely withheld from the model tuning and training process and used to assess the accuracy of the final model.

The PyHAT in-situ GUI includes a cross validation module that allows the user to evaluate multiple different regression algorithms and multiple different parameter settings for each algorithm to identify the optimal method and settings for the task at hand.

It also includes the capability to "blend" predictions from submodels trained on a restricted range of values. Model blending can be useful to improve the overall accuracy, particularly when handling data from diverse

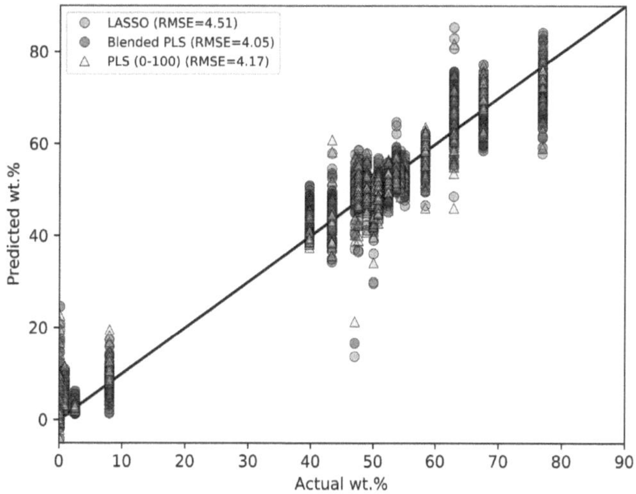

Figure 4.14 Predicted test set SiO_2 wt.% vs. actual SiO_2 wt.%. The blended PLS models are more accurate overall (lower RMSE) than either PLS or LASSO alone.

targets. To blend submodels, the cross validation process is repeated and the optimal settings for each submodel are found. An initial model spanning the full range of values is used as a first guess, which in turn determines which submodel(s) to use for the final prediction. The blended submodel approach is a simple method that has some of the same advantages as ensemble methods. For more information about the submodel regression approach, refer to Anderson et al. [2].

The PyHAT in-situ GUI also includes the capability to plot regression results, including automatic calculation of the Root Mean Squared Error (RMSE), the standard measure of a regression model's accuracy. Fig. 4.14 shows the test-set results of predicting SiO_2 wt.% based on LIBS spectra. Optimized Partial Least Squares (PLS) and Least Absolute Shrinkage and Selection Operator (LASSO) model results are compared with the results of blending several PLS submodels. The blended submodels have a lower overall RMSE, indicating greater accuracy.

4.4.3 Data exploration example

In this example, we demonstrate the PyHAT in-situ GUI's data exploration capabilities within Jupyter Notebook. Once the data have been loaded, masked, and normalized (Fig. 4.15), we perform dimensionality reduction. The GUI allows the user to choose between PCA [54], Locally Linear Embedding (LLE) [40], t-distributed Stochastic Neighbor Embedding (t-SNE) [49], and two different Independent Component Analysis (ICA) methods: "Fast ICA" and "Joint Approximation Diagonalization of Eigen-matrices (JADE) ICA" [8], [21]. For this example, we use PCA and t-SNE.

```
In [1]: import libpyhat
        from libpyhat.transform import mask, norm, dim_red
        import libpyhat.clustering.cluster as cluster
        import pandas as pd
        import matplotlib.pyplot as plot
        import numpy as np
```

```
In [2]: #Load the data
        datafile = "JSC_data_combined_20170307.csv"
        data = pd.read_csv(datafile,header=[0,1],low_memory=False)
```

```
In [3]: #Mask and Normalize the data
        maskfile = "mask_noise.csv"
        data = mask.mask(data,maskfile)
        data = norm.norm(data,[[0,350],[351,470],[471,850]])
```

```
In [4]: #Run PCA and get data for plotting
        data, dim_red_obj = dim_red.dim_red(data,'wvl','PCA',[],{'n_components':15})
        loading1 = dim_red_obj.components_[0,:]
        loading2 = dim_red_obj.components_[1,:]
        var1 = dim_red_obj.explained_variance_ratio_[0]*100
        var2 = dim_red_obj.explained_variance_ratio_[1]*100
        x = data[('PCA','PCA-1')]
        y = data[('PCA','PCA-2')]
        xlabel = 'PC 1 ('+str(round(var1,2))+'%)'
        ylabel = 'PC 2 ('+str(round(var2,2))+'%)'
        colorvar = 'MgO'
        wvls = data['wvl'].columns.values
```

Figure 4.15 Loading and preprocessing data for dimensionality reduction in PyHAT using Jupyter Notebook.

PCA is a widely used and relatively simple algorithm that maximizes the variance along orthogonal axes through the high dimensional data cloud, where each spectrum is a point in the cloud and each spectral channel is one dimension. To visualize the PCA results, we use one of the built-in plotting modules in the GUI, called "Plot ICA/PCA." This module creates plots showing the scores and loadings that result from ICA or PCA dimensionality reduction. Scores are the values of the spectra when projected onto the new axes defined by the algorithm, and loadings are the weight vectors which are multiplied by the spectra to derive the scores. Inspecting the loadings alongside the scores can help interpret the physical significance of the scores. Color coding the points in the scores plot can also be informative. For example, in Fig. 4.16, the PCA scores plot is color coded by the known abundance of MgO in the targets. There is a clear

In [5]:
```
#Make the PCA plot
fig = plot.figure()
fig.set_size_inches(10, 4)
ax1 = fig.add_subplot(2, 2, (1, 3))
ax2 = fig.add_subplot(2, 2, 2)
ax3 = fig.add_subplot(2, 2, 4, xlabel='Wavelength (nm)')
ax1.set_xlabel(xlabel)
ax1.set_ylabel(ylabel)
mappable = ax1.scatter(np.squeeze(x), np.squeeze(y), c=data[('comp', 'MgO')], cmap='viridis',
                       linewidth=0.2, edgecolor='Black')
fig.colorbar(mappable, label=colorvar, ax=ax1)
ax2.plot(wvls, loading1, linewidth=0.5)
ax3.plot(wvls, loading2, linewidth=0.5)
ax2.set_yticklabels([])
ax2.set_xticklabels([])
ax2.set_ylabel(xlabel)
ax3.set_yticklabels([])
ax3.set_ylabel(ylabel)
fig.savefig('PCA_fig.png', dpi=1000)
```

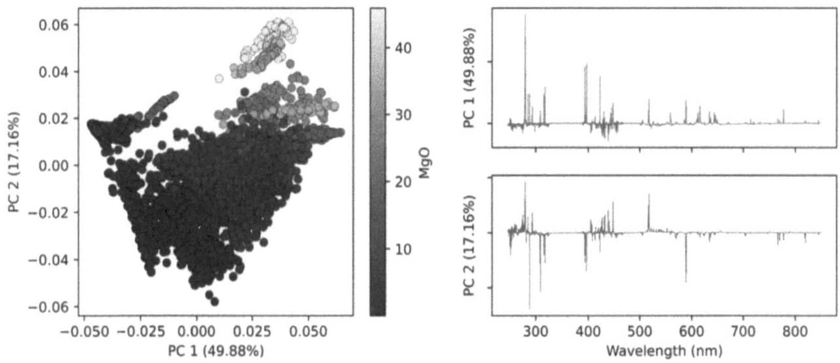

Figure 4.16 Scores (left) and loadings (right) for the first two principal components (with % variance explained in parentheses on axis labels) resulting from applying PCA to the masked and normalized data. Points on the scores plot are colored based on the known MgO wt.% of the targets. Loadings are the weight vectors multiplied by the spectra to derive the scores. The plots on the right reveal that both components have a strong positive weight around the Mg emission lines at ~280 nm, explaining why the highest MgO points plot in the upper-right of the scores plot. Note that the scores and loadings are unitless. Also note that to replicate the built-in plotting modules of the GUI in Jupyter Notebook, we have copied commands from the module into the notebook.

trend showing that high MgO spectra have scores that are positive in the first two principal components. Inspecting the loadings plot shows why: loadings for the first two components have positive weights corresponding to MgO emission lines (e.g., near 280 nm).

Whereas PCA performs linear dimensionality reduction, t-SNE (van der Maaten and Hinton [49]) is a nonlinear manifold-based method that is well-suited for visualizing the structure of high-dimensionality data in 2 or 3 dimensions. The outputs of t-SNE can be easier to visualize than other dimensionality reduction methods because it allows for the visualization of

```
In [6]: #Run k-means clustering on PCA scores
        params = {'n_clusters': 6, 'n_init': 10}
        data = cluster.cluster(data, 'PCA', 'K-Means', [], params)

In [7]: #run tSNE
        tsne_params = {'n_components': 2, 'learning_rate': 200.0,'perplexity': 175}
        data, tsne_obj = dim_red.dim_red(data,'PCA','t-SNE',[],tsne_params)

In [8]: fig = plot.figure()
        ax = fig.add_subplot(1, 1, 1)
        mappable = ax.scatter(data[('t-SNE','t-SNE-1')], data[('t-SNE','t-SNE-2')],
                              c=data[('K-Means', 'K-Means-Cluster')], cmap='viridis', linewidth=0.2, edgecolor='Black')
        fig.colorbar(mappable, label='K-Means Cluster', ax=ax)
        ax.set_xticklabels([])
        ax.set_yticklabels([])
        fig.savefig('tSNE_kmeans_fig.png', dpi=1000)
```

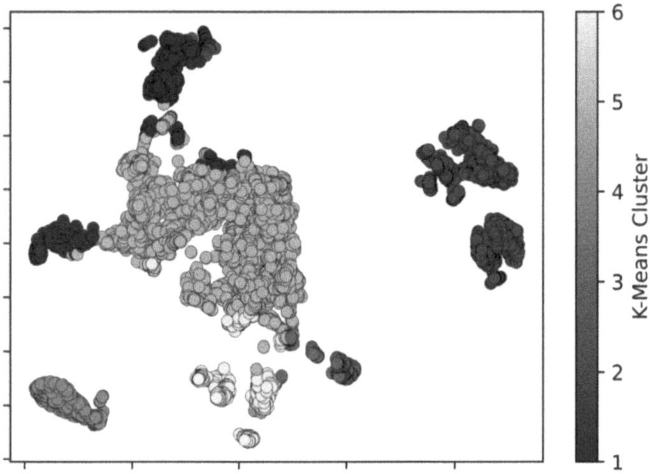

Figure 4.17 t-SNE plot of the same data as in Fig. 4.16, revealing structure that is not obvious in the PCA scores plot. Colors correspond to k-means clusters and illustrate how t-SNE distorts distances to better visualize the data (e.g., the blue (mid gray in print version) points on the edge of the purple (dark gray in print version) cluster are close to the rest of the purple (dark gray in print version) cluster in the PCA space used for k-means, but not in the 2D t-SNE map).

data points that are very close together, and thus similar, as well as those that are very dissimilar. PyHAT uses the scikit-learn implementation of t-SNE. Fig. 4.17 shows a t-SNE representation of the spectral data, which brings out structure that is not obvious in the PCA scores plot. K-means was run on the PCA scores for 15 components, which captures almost all of the variance in the spectral data.

The PyHAT in-situ GUI currently implements k-means [3] and spectral clustering [51] from scikit-learn, though additional clustering algorithms will be added in the future. For this example we ran k-means clustering on the PCA scores, with the number of clusters set to six. The t-SNE plot in

Fig. 4.17 has been color coded based on these clusters. Note that some of the colored points are plotted in a location that appears to be distinct from the main cluster of that color. This is due to the distortion of the high-dimensionality data that is performed by t-SNE. Those points are close to each other in the high-dimensional space, but not on the two dimensional (2D) mapping.

4.4.4 Calibration transfer

We next demonstrate the calibration transfer capabilities of the PyHAT in-situ GUI. We have already loaded and preprocessed one data set as described above, and now we load a separate data set collected on a different instrument on a subset of the samples. The second data set is preprocessed in the same way by masking and normalizing. Then the first data set is resampled to the same spectral channels as the second data set so that the spectra are precisely compatible. After that, we run calibration transfer cross validation. There are 11 different calibration transfer algorithms available in PyHAT, each of which may have several parameters to tune. The cross-validation module allows the user to efficiently iterate over different algorithms and parameter settings to determine the best choice. For simplicity in this example, we use piecewise direct standardization [52]. This algorithm is relatively fast and effective for LIBS spectra.

The calibration transfer cross validation module takes two data sets as inputs and identifies the spectra that they have in common based on a specified matching column. In this example we match on the sample name. Repeated spectra with the same sample name in each data set are averaged together so that the module ends up with two data sets, each with a single average spectrum for each sample that they have in common. At this point the cross validation begins.

For each permutation of model settings, each sample is iteratively held out from the data sets, a calibration transfer model is trained on the remaining spectra, and then applied to the held-out spectrum. The performance of the calibration transfer is calculated based on the difference between the transformed spectrum and the spectrum from the data set that the transform is supposed to match. The average RMSE after iterating over all of the spectra gives an idea of which algorithm and parameters give the best results.

For this example, we cross validate piecewise direct standardization using a window of between 1 and 9 spectral channels and find that a window of 7 channels gave the best result. Fig. 4.18 shows an example spectrum from

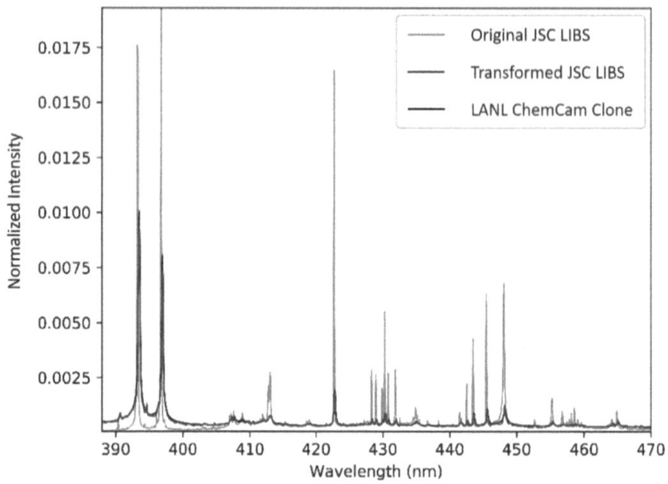

Figure 4.18 Example of calibration transfer using piecewise direct standardization and a window of seven spectral channels. The original JSC LIBS spectrum (red (mid gray in print version)) is transformed (blue (dark gray in print version)) to be very similar to the spectrum of the same target observed on a different instrument, the ChemCam clone at LANL (black, which is mostly overlapped by blue (dark gray in print version) due to the strong match).

the first data set before (red (mid gray in print version)) and after (blue (dark gray in print version)) applying piecewise direct standardization, compared with the target spectrum in the second data set (black).

4.5 Conclusion

PyHAT is a hyperspectral data analysis tool composed of a low-level API suitable for the development of more complex analysis and environments. It can also be used for exploratory data analysis inside of Jupyter notebooks, or in two GUIs (one for orbital data, one for in-situ data) designed to support nondeveloper scientists who seek a more user friendly experience. To demonstrate the uses of the PyHAT library, we have presented both orbital and in-situ applications of the software.

The derived products presented here also highlight our efforts to ensure that the equations being frequently used and sometimes published by mission science teams can be readily used by nonteam members, their results can be accurately reproduced, and the equations can be edited as needed for further exploratory research. We have demonstrated the efficacy of the

QGIS/PyHAT orbital plugin for creating derived products for quick-look mineral surveys on the Moon and Mars with M^3 and CRISM data, respectively. This ability to freely reproduce data derivation pipelines for hyper- and multispectral data is of great value to the planetary science community. Similarly, we hope that the demonstrated application of spectral analysis methods within the Jupyter notebook environment is a powerful motivator for those research scientists interested in a more interactive, exploratory data visualization and analysis environment that bridges the gap between a traditional GUI and more involved Python programming.

PyHAT orbital and PyHAT in-situ and the associated GUIs are available on GitHub and are hosted under a U.S. Geological Survey Astrogeology Science Center organizational page. Installation instructions and documentation are hosted adjacent to these software repositories, and a user can track changes as capabilities improve. Precompiled, cross-platform binaries are available using the Anaconda Python environment and installation is straightforward using these standard package management solutions. We are seeking to remove barriers for usage in order to make PyHAT as user-friendly and easy to adopt as possible.

We believe that PyHAT is a valuable addition to the free and open-source corpus of software that supports the planetary sciences. Currently funded efforts in PyHAT in-situ include the implementation of additional algorithms for dimensionality reduction, classification, end-member identification, and spectral unmixing. This work draws from algorithms in the terrestrial remote sensing and machine learning literature, and will be made available first in the PyHAT in-situ GUI, then in the PyHAT orbital GUI. The PyHAT in-situ GUI is already in use by the ChemCam and SuperCam science teams, and additional capabilities, both in the underlying PyHAT library and the PyHAT in-situ GUI, may be implemented in response to the needs of those teams. Future work may also involve revising the underlying data structures used in PyHAT such that the orbital and in-situ methods work with a common data structure.

Future work may include minor refactoring to the underlying API to improve user experience and to handle issues with copying versus slicing Pandas Data Frames. Copies versus slices are a known point of complexity of the current implementation and one we are excited to complete work on. Additionally, we may add support for spectral extraction and comparison of results to spectral libraries and the analytical methods that would better support comparison of data to well-known laboratory spectral collections. Spectral library capabilities have been prototyped and many of the components to accomplish this effort are already in place.

Acronyms

API	application programming interface
CAT	CRISM Analysis Toolkit
Ch1	Chandrayaan-1
CRISM	Compact Reconnaissance Imaging Spectrometer
CSV	Comma Separated Value
DN	Digital Number
ENVI	Environment for Visualizing Images
FAL	FALse-color infrared
FoV	Field of View
FRT	Full Resolution Targets
GDAL	Geospatial Data Abstraction Library
GIS	geographic information system
GM	Global Mode
GUI	graphical user interface
HCP	High-Calcium Pyroxene
HRT	Half Resolution Targets
ICA	Independent Component Analysis
IDL	Interactive Data Language
IRA	InfraRed Albedo
ISRO	Indian Space Research Organization
JADE	Joint Approximation Diagonalization of Eigen-matrices
JAXA	Japan Aerospace Exploration Agency
JSC	Johnson Space Center
KK	Kajfosz-Kwiatek
LANL	Los Alamos National Laboratory
LASSO	Least Absolute Shrinkage and Selection Operator
LCP	Low-Calcium Pyroxene
LIBS	Laser-Induced Breakdown Spectroscopy
LLE	Locally Linear Embedding
M^3	Moon Mineralogy Mapper
MAF	MAFic mineralogy
MTRDR	Map-Projected Targeted Reduced Data Records
MRO	Mars Reconnaissance Orbiter
NASA	National Aeronautics and Space Administration
NIR	Near-InfraRed
OP	Optical Periods
PCA	Principal Component Analysis
PDS	Planetary Data System
PLS	Partial Least Squares
PyHAT	Python Hyperspectral Analysis Tool
QGIS	Quantum Geographic Information System
RGB	Red Green Blue

RMSE	Root Mean Squared Error
SELENE	SELenological and ENgineering Explorer
SP	Spectral Profiler
SQL	Structured Query Language
TM	Target Mode
t-SNE	t-distributed Stochastic Neighbor Embedding
VIS	Visible

Acknowledgments

This work has been supported by NASA-USGS Inter-agency Agreement #NNH16AC13I and PDART agreements #NNH15AZ89I and #NNH18ZDA001N. Any use of trade, firm, or product names is for descriptive purposes only and does not imply endorsement by the U.S. Government.

References

[1] J.B. Adams, Visible and near-infrared diffuse reflectance spectra of pyroxenes as applied to remote sensing of solid objects in the solar system, Journal of Geophysical Research (1896-1977) 79 (32) (1974) 4829–4836, https://doi.org/10.1029/JB079i032p04829.

[2] R. Anderson, S. Clegg, J. Frydenvang, R. Wiens, S. McLennan, R. Morris, et al., Improved accuracy in quantitative laser-induced breakdown spectroscopy using sub-models, Spectrochimica Acta Part B 129 (2017) 49–57, https://doi.org/10.1016/j.sab.2016.12.002.

[3] D. Arthur, S. Vassilvitskii, k-means++: the advantages of careful seeding, in: SODA'07: Proceedings of the Eighteenth Annual ACM-SIAM Symposium on Discrete Algorithms, 2007, pp. 1027–1035.

[4] L. Barba, N. Clementi, G. Forsyth, The Python scientific stack, Web Page, retrieved 8/6/2020, from https://barbagroup.github.io/essential_skills_RRC/jupyter/1/, 2016.

[5] S. Besse, J.M. Sunshine, M.I. Staid, N.E. Petro, J.W. Boardman, R.O. Green, et al., Compositional variability of the Marius hills volcanic complex from the moon mineralogy mapper (m3), Journal of Geophysical Research: Planets 116 (E6) (2011), https://doi.org/10.1029/2010JE003725.

[6] J.W. Boardman, C.M. Pieters, R.O. Green, S.R. Lundeen, P. Varanasi, J. Nettles, et al., Measuring moonlight: an overview of the spatial properties, lunar coverage, selenolocation, and related level 1b products of the moon mineralogy mapper, Journal of Geophysical Research: Planets 116 (E6) (2011), https://doi.org/10.1029/2010JE003730.

[7] R. Burns, M. Hochella, R. Liebermann, A. Putnis, Mineralogical Applications of Crystal Field Theory, Cambridge University Press, 1993, retrieved from https://books.google.com/books?id=izKuGT5Bjp8C.

[8] J. Cardoso, A. Souloumiac, Blind beamforming for non-Gaussian signals, IEE Proceedings 6 (1993).

[9] L.C. Cheek, C.M. Pieters, J.W. Boardman, R.N. Clark, J.P. Combe, J.W. Head, et al., Goldschmidt crater and the moon's North polar region: results from the moon mineralogy mapper (m3), Journal of Geophysical Research: Planets 116 (E6) (2011), https://doi.org/10.1029/2010JE003702.

[10] R.N. Clark, T.V.V. King, M. Klejwa, G.A. Swayze, N. Vergo, High spectral resolution reflectance spectroscopy of minerals, Journal of Geophysical Research: Solid Earth 95 (B8) (1990) 12653–12680, https://doi.org/10.1029/JB095iB08p12653.

[11] E.A. Cloutis, M.J. Gaffey, Spectral-compositional variations in the constituent minerals of mafic and ultramafic assemblages and remote sensing implications, Earth, Moon, and Planets 53 (1) (1991) 11–53, https://doi.org/10.1007/BF00116217.

[12] E.A. Cloutis, F.C. Hawthorne, S.A. Mertzman, K. Krenn, M.A. Craig, D. Marcino, et al., Detection and discrimination of sulfate minerals using reflectance spectroscopy, Icarus 184 (1) (2006) 121–157, https://doi.org/10.1016/j.icarus.2006.04.003.

[13] S. Giguere, C. Carey, M. Dyar, T. Boucher, M. Parente, T. Tague Jr, S. Mahadevan, Baseline removal in LIBS and FTIR spectroscopy: optimization techniques, in: 46th Lunar and Planetary Science Conference, 2015, p. 2.

[14] J. Goswami, M. Annadurai, Chandrayaan-1: India's first planetary science mission to the moon, Current Science 96 (2009, 01).

[15] R.O. Green, C. Pieters, P. Mouroulis, M. Eastwood, J. Boardman, T. Glavich, et al., The moon mineralogy mapper (m3) imaging spectrometer for lunar science: instrument description, calibration, on-orbit measurements, science data calibration and on-orbit validation, Journal of Geophysical Research: Planets 116 (E10) (2011), https://doi.org/10.1029/2011JE003797.

[16] J. Haruyama, T. Matsunaga, M. Ohtake, T. Morota, C. Honda, Y. Yokota, et al., Global lunar-surface mapping experiment using the lunar imager/spectrometer on SELENE, Earth, Planets and Space 60 (4) (2008) 243–255, https://doi.org/10.1186/BF03352788.

[17] B.H. Horgan, R.B. Anderson, G. Dromart, E.S. Amador, M.S. Rice, The mineral diversity of Jezero crater: evidence for possible lacustrine carbonates on Mars, Icarus 339 (2020) 113526, https://doi.org/10.1016/j.icarus.2019.113526.

[18] B.H. Horgan, E.A. Cloutis, P. Mann, J.F. Bell III, Near-infrared spectra of ferrous mineral mixtures and methods for their identification in planetary surface spectra, Icarus 234 (2014) 132–154, https://doi.org/10.1016/j.icarus.2014.02.031.

[19] G.R. Hunt, J.W. Salisbury, Visible and infrared spectra of minerals and rocks. IV: sulphides and sulphates, Modern Geology 3 (1971, January) 1–14.

[20] G.R. Hunt, J.W. Salisbury, Visible and near infrared spectra of minerals and rocks. II. Carbonates, Modern Geology 2 (1971, January) 23–30.

[21] A. Hyvarinen, E. Oja, Independent component analysis: algorithms and applications, Neural Networks 13 (1993) 411–430, https://doi.org/10.1016/S0893-6080(00)00026-5.

[22] P.J. Isaacson, C.M. Pieters, S. Besse, R.N. Clark, J.W. Head, R.L. Klima, et al., Remote compositional analysis of lunar olivine-rich lithologies with moon mineralogy mapper (m3) spectra, Journal of Geophysical Research: Planets 116 (E6) (2011), https://doi.org/10.1029/2010JE003731.

[23] J. Kajfosz, W. Kwiatek, Nonpolynomial approximation of background in x-ray spectra, Nuclear Instruments and Methods in Physics Research B (1987) 78–81.

[24] R. Klima, J. Cahill, J. Hagerty, D. Lawrence, Remote detection of magmatic water in Bullialdus crater on the moon, Nature Geoscience 6 (9) (2013, 09) 737–741, https://doi.org/10.1038/ngeo1909.

[25] T. Kluyver, B. Ragan-Kelley, F. Perez, B. Granger, M. Bussonnier, J. Frederic, et al., Jupyter notebooks – a publishing format for reproducible computational workflows, in: 20th International Conference on Electronic Publishing, 2016, 01.

[26] G.Y. Kramer, S. Besse, D. Dhingra, J. Nettles, R. Klima, I. Garrick-Bethell, et al., M3 spectral analysis of lunar swirls and the link between optical maturation and surface hydroxyl formation at magnetic anomalies, Journal of Geophysical Research: Planets 116 (E9) (2011), https://doi.org/10.1029/2010JE003729.

[27] C. Lieber, A. Mahadevan-Jansen, Automated method for subtraction of fluorescence from biological Raman spectra, Applied Spectroscopy 57 (11) (2003) 1363–1367.

[28] P. Lucey, D. Blewett, B. Jolliff, Lunar iron and titanium abundance algorithms based on final processing of clementine ultraviolet-visible images, Journal of Geophysical Research 105 (2000) 20297–20305.
[29] T. Matsunaga, M. Ohtake, J. Haruyama, Y. Ogawa, R. Nakamura, Y. Yokota, et al., Discoveries on the lithology of lunar crater central peaks by SELENE spectral profiler, Geophysical Research Letters 35 (23) (2008) L23201, https://doi.org/10.1029/2008GL035868.
[30] T.B. McCord, L.A. Taylor, J.-P. Combe, G. Kramer, C.M. Pieters, J.M. Sunshine, R.N. Clark, Sources and physical processes responsible for oh/h2o in the lunar soil as revealed by the moon mineralogy mapper (m3), Journal of Geophysical Research: Planets 116 (E6) (2011), https://doi.org/10.1029/2010JE003711.
[31] F. Morgan, F.P. Seelos, S.L. Murchie, CRISM analysis toolkit (CAT), in: L.R. Gaddis, T.M. Hare, R. Beyer (Eds.), Summary and Abstracts of the Planetary Data Workshop, June 2012, 2014, pp. 125–126.
[32] S. Murchie, R. Arvidson, P. Bedini, K. Beisser, J.-P. Bibring, J. Bishop, et al., Compact reconnaissance imaging spectrometer for Mars (CRISM) on Mars reconnaissance orbiter (MRO), Journal of Geophysical Research: Planets 112 (E5) (2007), https://doi.org/10.1029/2006JE002682.
[33] S.L. Murchie, F.P. Seelos, C.D. Hash, D.C. Humm, E. Malaret, J.A. McGovern, et al., Compact reconnaissance imaging spectrometer for Mars investigation and data set from the Mars reconnaissance orbiter's primary science phase, Journal of Geophysical Research: Planets 114 (E2) (2009), https://doi.org/10.1029/2009JE003344.
[34] J.F. Mustard, C.M. Pieters, P.J. Isaacson, J.W. Head, S. Besse, R.N. Clark, et al., Compositional diversity and geologic insights of the Aristarchus crater from moon mineralogy mapper data, Journal of Geophysical Research: Planets 116 (E6) (2011), https://doi.org/10.1029/2010JE003726.
[35] J.W. Nettles, M. Staid, S. Besse, J. Boardman, R.N. Clark, D. Dhingra, et al., Optical maturity variation in lunar spectra as measured by moon mineralogy mapper data, Journal of Geophysical Research: Planets 116 (E9) (2011), https://doi.org/10.1029/2010JE003748.
[36] S.M. Pelkey, J.F. Mustard, S. Murchie, R.T. Clancy, M. Wolff, M. Smith, et al., CRISM multispectral summary products: parameterizing mineral diversity on Mars from reflectance, Journal of Geophysical Research: Planets 112 (E8) (2007), https://doi.org/10.1029/2006JE002831.
[37] C.M. Pieters, S. Besse, J. Boardman, B. Buratti, L. Cheek, R.N. Clark, et al., Mg-spinel lithology: a new rock type on the lunar farside, Journal of Geophysical Research: Planets 116 (E6) (2011), https://doi.org/10.1029/2010JE003727.
[38] C.M. Pieters, J.N. Goswami, R.N. Clark, M. Annadurai, J. Boardman, B. Buratti, et al., Character and spatial distribution of oh/h2o on the surface of the moon seen by m3 on Chandrayaan-1, Science 326 (5952) (2009) 568–572, https://doi.org/10.1126/science.1178658.
[39] G.R. Rossman, Spectroscopic and magnetic studies of ferric iron hydroxy sulfates: the series fe(oh)$so_4 \cdot nh_2o$ and the jarosites, American Mineralogist 61 (506) (1976) 398–404.
[40] S. Roweis, L. Saul, Nonlinear dimensionality reduction by locally linear embedding, Science 290 (2000) 2323–2326.
[41] S. Sasaki, Y. Iijima, K. Tanaka, M. Kato, M. Hashimoto, H. Mizutani, Y. Takizawa, The moon: science, exploration and utilisation the SELENE mission: goals and status, Advances in Space Research 31 (11) (2003) 2335–2340, https://doi.org/10.1016/S0273-1177(03)00543-X.
[42] F.P. Seelos, S.L. Murchie, D.C. Humm, O.S. Barnouin, F. Morgan, H.W. Taylor, et al., CRISM Team, CRISM data processing and analysis products update — calibra-

tion, correction, and visualization, in: Lunar and Planetary Science Conference, 2011, March, p. 1438.
[43] D.M. Sherman, R.G. Burns, V.M. Burns, Spectral characteristics of the iron oxides with application to the Martian bright region mineralogy, Journal of Geophysical Research: Solid Earth 87 (B12) (1982) 10169–10180, https://doi.org/10.1029/JB087iB12p10169.
[44] M.I. Staid, C.M. Pieters, S. Besse, J. Boardman, D. Dhingra, R. Green, et al., The mineralogy of late stage lunar volcanism as observed by the moon mineralogy mapper on Chandrayaan-1, Journal of Geophysical Research: Planets 116 (E6) (2011), https://doi.org/10.1029/2010JE003735.
[45] J.M. Sunshine, C.M. Pieters, Determining the composition of olivine from reflectance spectroscopy, Journal of Geophysical Research: Planets 103 (E6) (1998) 13675–13688, https://doi.org/10.1029/98JE01217.
[46] J.M. Sunshine, C.M. Pieters, S.F. Pratt, Deconvolution of mineral absorption bands: an improved approach, Journal of Geophysical Research: Solid Earth 95 (B5) (1990) 6955–6966, https://doi.org/10.1029/JB095iB05p06955.
[47] K.G. Thaisen, J.W. Head, L.A. Taylor, G.Y. Kramer, P. Isaacson, J. Nettles, et al., Geology of the moscoviense basin, Journal of Geophysical Research: Planets 116 (E6) (2011), https://doi.org/10.1029/2010JE003732.
[48] J.W. Tukey, Exploratory Data Analysis, Addison-Wesley, 1977.
[49] L. van der Maaten, G. Hinton, Visualizing data using t-sne, Journal of Machine Learning Research 9 (2008) 2579–2605.
[50] C.E. Viviano-Beck, F.P. Seelos, S.L. Murchie, E.G. Kahn, K.D. Seelos, H.W. Taylor, et al., Revised CRISM spectral parameters and summary products based on the currently detected mineral diversity on Mars, Journal of Geophysical Research: Planets 119 (6) (2014) 1403–1431, https://doi.org/10.1002/2014JE004627.
[51] U. von Luxburg, A tutorial on spectral clustering, CoRR, arXiv:0711.0189, 2007.
[52] Y. Wang, D. Veltkamp, B. Kowalski, Multivariate instrument standardization, Analytical Chemistry 63 (1991) 2750–2756, https://doi.org/10.1021/ac00023a016.
[53] J. Whitten, J.W. Head, M. Staid, C.M. Pieters, J. Mustard, R. Clark, et al., Lunar mare deposits associated with the orientale impact basin: new insights into mineralogy, history, mode of emplacement, and relation to orientale basin evolution from moon mineralogy mapper (m3) data from Chandrayaan-1, Journal of Geophysical Research: Planets 116 (E6) (2011), https://doi.org/10.1029/2010JE003736.
[54] S. Wold, K. Esbensen, P. Geladi, Principal component analysis, Chemometrics and Intelligent Laboratory Systems 2 (1987) 37–52.
[55] S. Yamamoto, T. Matsunaga, Y. Ogawa, R. Nakamura, Y. Yokota, M. Ohtake, et al., Preflight and in-flight calibration of the spectral profiler on board SELENE (Kaguya), IEEE Transactions on Geoscience and Remote Sensing 49 (11) (2011, Nov) 4660–4676, https://doi.org/10.1109/TGRS.2011.2144990.
[56] S. Yamamoto, T. Matsunaga, Y. Ogawa, R. Nakamura, Y. Yokota, M. Ohtake, et al., Calibration of NIR 2 of spectral profiler onboard Kaguya/SELENE, IEEE Transactions on Geoscience and Remote Sensing 52 (11) (2014, Nov) 6882–6898, https://doi.org/10.1109/TGRS.2014.2304581.
[57] R.W. Zurek, S.E. Smrekar, An overview of the Mars reconnaissance orbiter (MRO) science mission, Journal of Geophysical Research: Planets 112 (E5) (2007), https://doi.org/10.1029/2006JE002701.

CHAPTER 5

Tutorial: how to access, process, and label PDS image data for machine learning☆

Steven Lu, Kiri L. Wagstaff, Rafael Alanis, Gary Doran, Kevin Grimes, and Jordan Padams
Jet Propulsion Laboratory, California Institute of Technology, Pasadena, CA, United States

5.1 Introduction

Machine learning methods have been used to achieve new advances in image analysis and classification problems that range from genetics [1] to materials science [2] to crop yield prediction and other agricultural applications [3]. In addition to these terrestrial applications, there is growing interest in the application of the same methods to images collected on other planets and bodies in support of space exploration. However, it can be difficult to know where and how to access these images, and they are often stored in data formats that are unfamiliar to the machine learning community. This tutorial aims to facilitate machine learning projects that can make use of the more than 700 TB (and growing) of NASA planetary science images.

Images collected by NASA planetary science missions are curated by the Cartography and Imaging Sciences Discipline Node (Imaging Node) of the Planetary Data System (PDS). These holdings currently include more than 34 million products that span 22 missions, with targets that include the Moon, Mars, Mercury, Jupiter, Saturn, Venus, and more. Some example images are shown in Fig. 5.1.

The PDS provides data search and retrieval tools to assist users with finding and downloading the image products of interest. There are also several options for preprocessing the image products into image formats that are directly usable by machine learning systems. This tutorial introduces these tools along with step-by-step examples of how to use them. For training a machine learning classifier, there is a common need to obtain labels for

☆ Contribution prepared by the Contributor on behalf of JPL/Caltech.

Machine Learning for Planetary Science
https://doi.org/10.1016/B978-0-12-818721-0.00013-6
Copyright © 2022 Elsevier Inc.
All rights reserved.

(a) Moon (Lunar Orbiter 1, 1966) (b) Jupiter (Voyager 2, 1979) (c) Venus (Magellan, 1991)

(d) Saturn (Cassini, 2004) (e) Mercury (MESSENGER, 2015) (f) Mars (InSight, 2019)

Figure 5.1 Example images curated by the PDS Imaging Node, spanning more than five decades.

concepts or features of interest within the images. This is not a challenge specific to planetary science images, but for completeness this tutorial also includes suggestions and advice to facilitate the labeling process. Finally, we describe an existing project that has employed PDS image data as an example end-to-end machine learning investigation.

5.2 Access to PDS data products

This section introduces tools provided by the PDS Imaging Node to browse, search, and download planetary mission image data. The PDS Image Atlas[1] provides a visual browser that allows users to zero in on data of interest by applying filters based on image metadata and/or content. The PDS Data Portal[2] provides direct access to the data archive directories. The

[1] PDS Image Atlas: https://pds-imaging.jpl.nasa.gov/search.
[2] PDS Data Portal: https://pds-imaging.jpl.nasa.gov/portal.

PDS also provides an online interface called PDS Tool Registry[3] that enables search and discovery of tools, services, and APIs for working with data that adheres to the PDS standards. Tool providers can submit new software for inclusion in the registry for others to use.

At a higher level, the PDS Keyword Search provides the ability to search and access all planetary data sets across the entire PDS, not just the Imaging Node. These data sets include observations from the disciplines of atmospheres, geosciences, particles and fields, and radio science. To learn more, navigate to https://pds.nasa.gov/datasearch/keyword-search/.

5.2.1 PDS Image Atlas

The PDS Image Atlas provides unique capabilities for users to find data of interest. It allows users to specify filters based on product metadata (e.g., target, illumination, time of year) and content (e.g., contains craters) to quickly narrow down the search and then download the selected data products.

Fig. 5.2(a) shows the PDS Image Atlas starting page. A search box at the top of the page provides full-text searching of image metadata. In the upper right, the total number of image products is shown (34,185,087 as of October 2019). Pagination allows the user to step through the images in batches; by default, they are displayed in reverse chronological order, but this can be configured by the user. In this screenshot, the most recent images are from the InSight mission on the surface of Mars.

Users can apply filters to reduce the number of image results using the left-side panel. For example, to obtain images collected by the Mars Science Laboratory rover, using the Mastcam instrument, from a specified time range that contain drill holes, follow these steps to apply the relevant metadata and content-based filters:

- Navigate to the PDS Image Atlas at https://pds-imaging.jpl.nasa.gov/search.
- Apply metadata-based filters:
  - From the "Mission" tab, select "mars science laboratory".
  - From the "Instrument" tab, select "mastcam".
  - From the "Target" tab, select "mars".
  - From the "Product Type" tab, select "edr" (Experiment Data Record[4]).

[3] PDS Tool Registry: https://pds.nasa.gov/tools/tool-registry.

[4] Raw data at full instrument resolution.

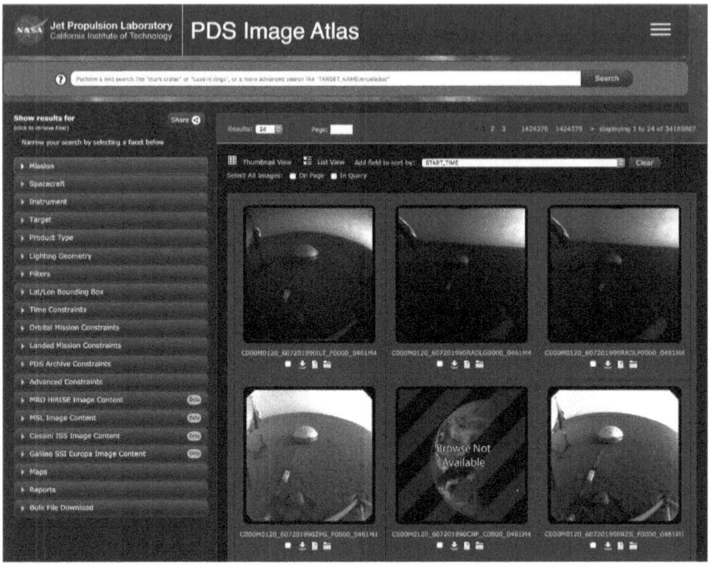

(a) PDS Image Atlas starting page (whole collection)

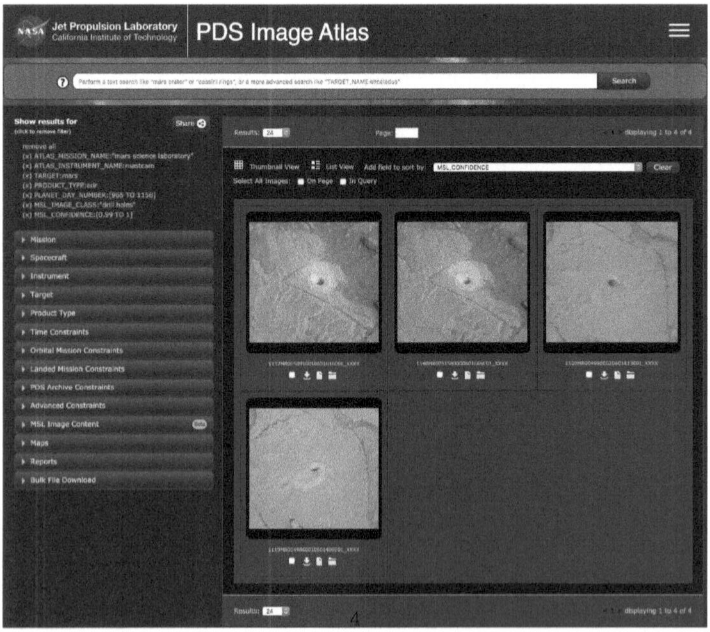

(b) PDS Image Atlas example after applying filters (see text). Note that the search results may be different because new images are continuously delivered to PDS.

Figure 5.2 The PDS Image Atlas.

  - From the "Landed Mission Constraints" tab, type 965 and 1158 in the "Planet Day Number" fields. Press enter.
- Apply content-based filters:
  - From the "MSL Image Content" tab, select "drill holes".
  - From the "MSL Image Content" tab, type 0.99 and 1.0 in the "Confidence Level" fields. Press enter.

The results are shown in Fig. 5.2(b). The 34 million images have been narrowed down to four. The relevant filters are shown in the upper left of the page, and they can be selectively removed to broaden the search as desired.

To download the images in the search results:

- Click the "Select All Images: In Query" checkbox.
- From the "Bulk File Download" tab, click "Begin Download" button.
- Execute the downloaded script to start retrieving the data products.

The Atlas is powered by Apache Solr,[5] an open source search platform built on Apache Lucene.[6] As the user interacts with the web interface, queries are sent to the Solr backend, and the interface is updated with the results.

Users and developers alike may choose to interact with the Solr application programming interface ("API") directly, rather than going through the Atlas's web interface. This enables users to make custom queries that cannot be made through the web interface. Additionally, developers may leverage the API in their applications by making HTTP requests to it. API documentation[7] for the Atlas is available online.

5.2.2 PDS Imaging Node Data Portal

The PDS Imaging Node Data Portal provides entry points to all planetary science image data sets released by missions and other data providers through the PDS Imaging Node. It allows users to easily navigate through different data volumes and connect to the back-end data services. For example, to download images collected by the Mars Reconnaissance Orbiter's HiRISE instrument using the PDS Imaging Node data portal:

- Navigate to the Data Portal site at https://pds-imaging.jpl.nasa.gov/portal/ (Fig. 5.3).

[5] Apache Solr: https://lucene.apache.org/solr.
[6] Apache Lucene: https://lucene.apache.org.
[7] PDS Image Atlas API: https://pds-imaging.jpl.nasa.gov/tools/atlas/api/.

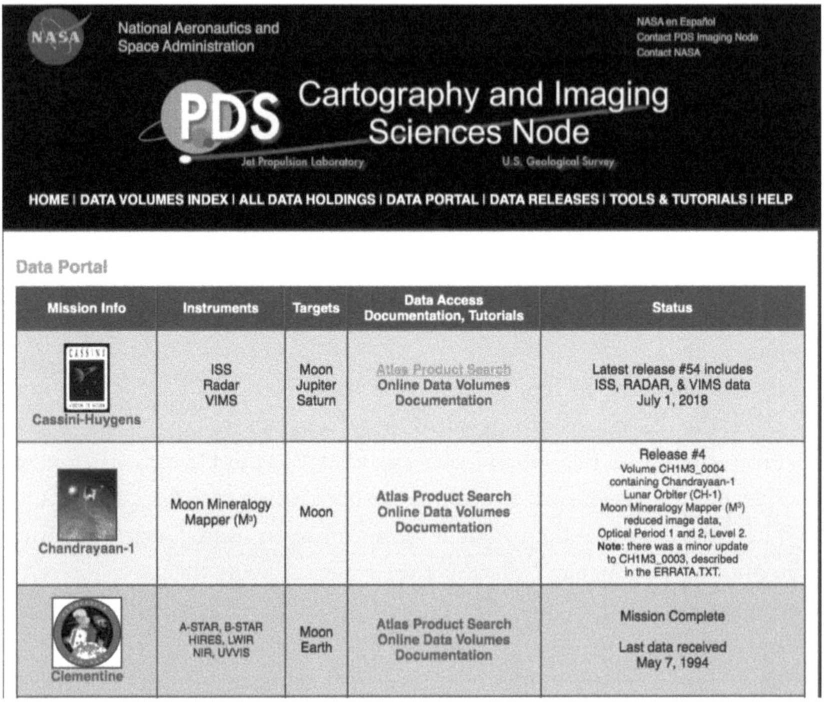

Figure 5.3 PDS Data Portal starting page.

- Scroll down the web page to find "Mars Reconnaissance Orbiter," and then click "Online Data Volumes" on the right.
- Click the dark red server icon next to the text "Volume 1 (accumulating)" in the first row of "Mission to Mars: High Resolution Imaging Science Experiment (HiRISE)" section. You will then be re-directed to the back-end data services.
- Navigate to the `/EXTRAS/RDR/ESP/ORB_061200_061299/ESP_061200_1790/` folder,[8] which contains images from orbit 61,200.
- Here, users may choose their favorite tools to download data products. Individual images can be viewed by clicking their filenames. The `wget` command[9] can be used to download (for example) all JPG products from this directory:

[8] The directory structure for data products varies by mission and instrument. See the `AAREADME.txt` file for a description of each instrument's chosen data directory structure.

[9] https://www.gnu.org/software/wget/.

```
wget -nd -r -P ./ -A jpg https://hirise-pds.lpl.arizona.
edu/PDS/EXTRAS/RDR/ESP/ORB_061200_061299/ESP_061200_1790/
```

Note that the URL should be all on one line when executed.

5.3 Preprocessing PDS data products into standard image formats

PDS image data exists in several different formats, as chosen by the mission that generated the data. This section focuses on methods for converting PDS image data into standard (8-bit) image formats for viewing and subsequent analysis by machine learning methods.

5.3.1 PDS image data products

In general, each PDS image product consists of two items: the data file and its associated metadata. This metadata includes information about the context in which the data was collected, such as local time, imaging target, illumination, etc. This information is often referred to as a "label" in PDS literature. In order to distinguish this information from the way "label" is used in a machine learning context, we will refer to the PDS data label as "metadata."

For decades, the PDS required planetary missions to use a standard called PDS3 [4] to describe how data (and metadata) should be structured in order to be archived. The image data is usually stored in array or raster form such as VICAR [5], PDS, or FITS [6] format. Many other types of data such as spectral cubes, tables, and text are also archived. PDS3 metadata is stored in Object Description Language (ODL) using key-value pairs that can be read by both humans and machines. The image data and metadata can be combined into a single file (referred to as "attached") or kept separate, but colocated ("detached"), with the same base name for both the data and metadata. However, the metadata file will have an extension of ".LBL".

In 2010, the PDS added support for the PDS4 standard [7], which stores metadata in XML format, easing the process of validation (e.g., checking that required values are present, values fall into valid ranges, adheres to XML formatting standard) and software parsing. Image data is stored as a two-dimensional pixel array. Images that span multiple wavelengths (including RGB images) may be stored in a single file or in multiple files (one per wavelength), as described in the PDS4 XML metadata label.

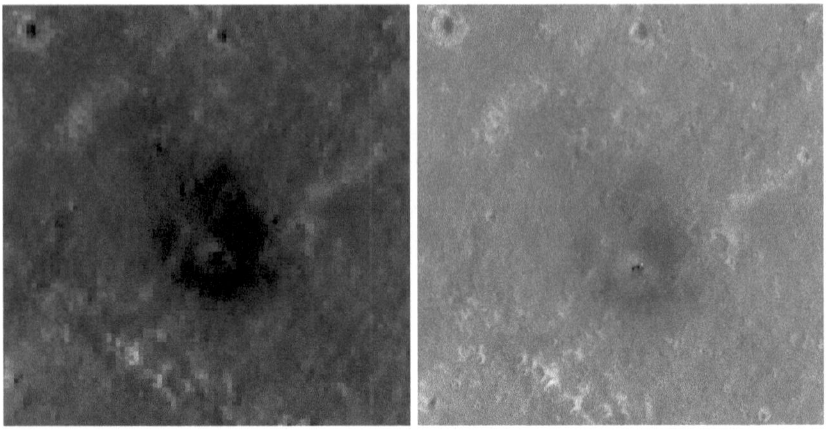

(a) HiRISE "browse" image of the InSight lander (the lander is not visible)

(b) HiRISE "JP2" image of the InSight lander (the lander is visible)

Figure 5.4 The "browse" and "JP2" versions of HiRISE product ESP_061684_1845. The InSight lander is visible only in the JP2 product to the right, and the bright spot near the center is the seismometer instrument.

In PDS4, the image data is stored similarly to PDS3, where VICAR, PDS, and FITS formats are often used, however, there are more refined, documented restrictions on the more detailed nuances of those formats as well as other less prevalent formats used in PDS3. Additionally, in PDS4, metadata for a data product must appear in an accompanying XML file ("detached"). More details about PDS4 data formats and all other information relating to the standard can be found in the PDS4 Data Provider's Handbook [8] and PDS4 Standards Reference [7]. The first mission to deliver image data to the PDS in the PDS4 format was the Mars InSight Lander.

5.3.2 PDS browse images

Many PDS data collections provide a "browse" version of the image products (commonly in JPG or PNG format) that enables fast preview of the contents. Browse images are not a science-quality product; they may employ lossy compression, automatic contrast stretch, and a reduction in bit depth. However, they can be useful for quick assessments. It is important to read the mission instrument documentation to understand the processing steps that were applied to generate the browse images and determine whether they will impact the planned analysis. Fig. 5.4 shows the difference between the "browse" and "JPEG 2000 (JP2)" versions of the HiRISE image product ESP_061684_1845.[10] Clearly, the InSight lander is only visible in Fig. 5.4(b), where the dark spots are the two circular solar panels and the bright spot is the seismometer instrument.

5.3.3 Converting PDS image data products

For the highest quality results, the best strategy is to access the PDS image product directly and convert it into the format needed. This enables use of the highest resolution (spatial and spectral) version of the data and control over any preprocessing that is applied.

Although the image formats employed for PDS image data are straightforward, the lack of a direct reader in popular and commonly used machine learning related packages (e.g., Python's Numpy, Scikit-learn) presents a challenge. One reason that NASA mission data is not natively stored in the more common 8-bit image formats such as GIF, JPG, and PNG is that the data is often collected at much higher bit depths (12, 16, or 32 bits) and therefore requires different formats. Here we introduce tools that can convert PDS into commonly used image formats, acknowledging that some detail may be lost.

- The **PDS Transform Tool** is a Java-based command-line software tool for transforming PDS3 and PDS4 products into other common formats (JPG, PNG, TIF, etc.). For example, to convert a Mars Science Laboratory image product (NRB_606845760RASLF0750750NCAM00592M1.IMG[11]) from PDS3 VICAR format to PNG format, using the accompanying PDS3 metadata label (NRB_606845760RASLF0750750NCAM00592M1.LBL[12]), the following command-line execution would generate a PNG file (Note: the software currently requires the IMG and LBL files to reside on the file system executing the software.):

```
transform -t NRB_606845760RASLF0750750NCAM00592M1.LBL
  -o NRB_606845760RASLF0750750NCAM00592M1 -f png
```

  Similar code could be used to convert a PDS4 VICAR formatted image to PNG format by providing the .xml as input in place of the .LBL. For more information on the installation and operation of the tool,

[10] https://www.uahirise.org/ESP_061684_1845.

[11] https://pds-imaging.jpl.nasa.gov/data/msl/MSLNAV_1XXX/DATA/SOL02358/NRB_606845760RASLF0750750NCAM00592M1.IMG.

[12] https://pds-imaging.jpl.nasa.gov/data/msl/MSLNAV_1XXX/DATA/SOL02358/NRB_606845760RASLF0750750NCAM00592M1.LBL.

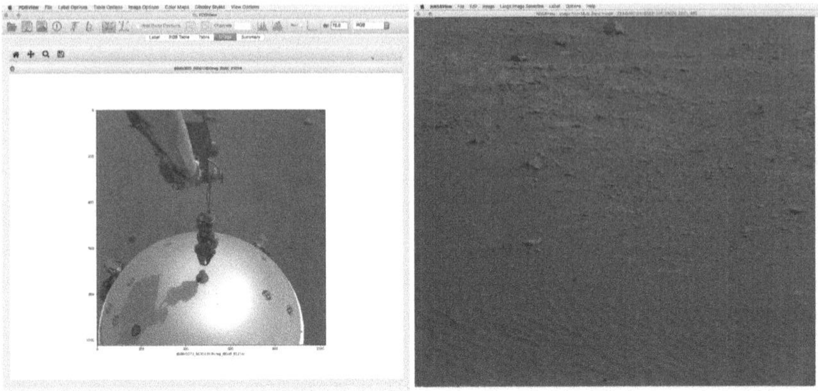

(a) (b)

Figure 5.5 (a) PDSView visualization of InSight image data. (b) NASAView visualization of MSL image data.

or to become a contributor to the software, go to https://nasa-pds-incubator.github.io/transform/.

- **IMG2PNG** is a utility which can convert images in IMG, VICAR, and FITS formats to PNG format. It is only available on the Windows platform. The installation instructions and parameters available for the tool can be found at http://bjj.mmedia.is/utils/img2png/.
- **PDSView** is a native Python application for visualizing PDS3 and PDS4 data products. It can be used to view metadata and image content, and it can also convert and save PDS data into PNG and other formats. Prebuilt installers are available for various versions of Windows, Macintosh, and Linux platforms. To find out more information, download the software, or become a contributor, go to https://github.com/NASA-PDS-Incubator/pds-view. Fig. 5.5(a) shows a screenshot of PDSView with an image collected by the InSight lander.
- **NASAView** is a PDS archive product display program that runs on multiple platforms (Solaris, Linux, Windows, Macintosh) in a GUI environment. The tool is capable of reading in images in IMG format, displaying them, and exporting them to GIF and JPG formats. For installation packages and detailed information, visit https://pds.nasa.gov/tools/about/pds3-tools/nasa-view.shtml. Fig. 5.5(b) shows a screenshot of NASAView with an image collected by the Mars Science Laboratory rover. This is a multiwavelength observation, visualized using a single band in grayscale.

- **USGS ISIS** (Integrated Software for Imagers and Spectrometers) is powerful image processing software package that can not only read PDS data but also perform calibration, reprojection, and other operations on the images. The focus of the software is to manipulate images collected by current and past NASA planetary missions sent to Mars, Jupiter, Saturn, and other solar system bodies. The **pds2isis** and **fits2isis** programs can read images in PDS3 or FITS format, respectively, into ISIS format, and then **isis2std** can convert the data into BMP, JPG, JP2, PNG, or TIF formats. More information is available at https://isis.astrogeology.usgs.gov/.

5.4 Labeling image data

Once the PDS image data has been converted into a standard image format (e.g., PNG or JPG), it can be labeled to provide the basis for training and evaluating a supervised machine learning model. In this section, we provide some suggestions for how to proceed as well as examples of existing PDS image data sets that are already labeled and available for use.

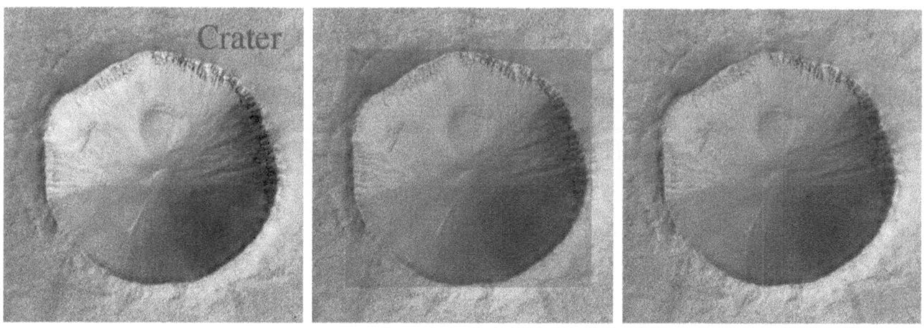

(a) Whole image labeling (b) Bounding box labeling (c) Polygon labeling

Figure 5.6 Examples of labeling tasks.

Image labeling tasks generally consist of three categories: (1) whole image labeling, (2) bounding box (rectangular subregion) labeling, and (3) polygon (complex region) labeling. Examples are shown in Fig. 5.6. Whole image labeling requires the labeler to assign a class or category to an entire image. Bounding box labeling requires the identification of regions of interest (bounding boxes) followed by the assignment of the relevant class. Polygon labeling employs more complex shapes to identify the regions of interest, which are often drawn by clicking the exterior points that define a polygon, followed by class assignment.

Different labeling strategies help achieve different classification goals. For example, if we would like to train a classifier to determine whether or not a Mars image contains craters, we need only perform whole image labeling. However, if we would like to train a classifier to be able to mark the precise locations of craters, we will need to perform polygon labeling. This tutorial will focus on whole image labeling in the examples.

5.4.1 Publicly available labeled image data sets

Some PDS image data has already been processed, converted, and labeled for use in machine learning investigations. Examples include the Mars Orbital Image, Mars Surface Image, and Planetary Body Limb and Plume data sets.

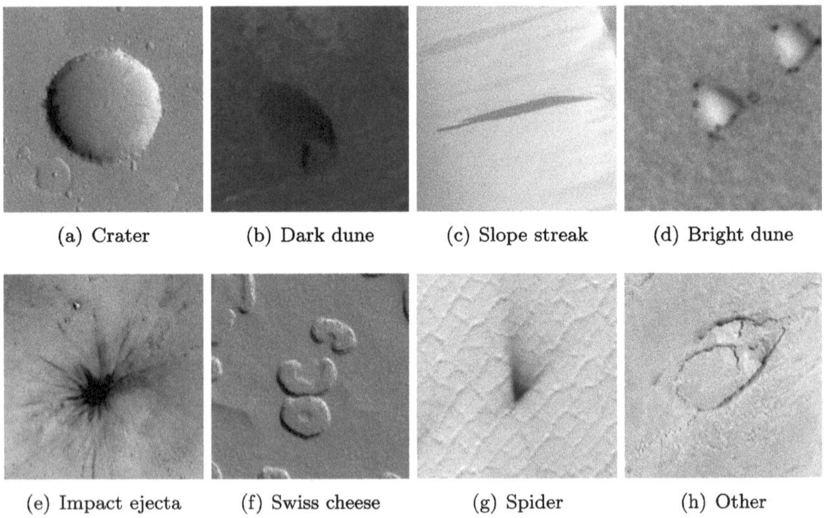

Figure 5.7 Examples of Mars Orbital Image (HiRISE) data set (version 3).

The **Mars Orbital Image (HiRISE)** data set (version 3) [9] contains "landmark" images cropped from HiRISE (orbital) map-projected browse images. Landmarks are visually salient surface features, such as craters, dunes, and dark slope streaks. The data set contains 73,031 landmark images that were obtained by detecting 10,433 landmark images from 180 HiRISE map-projected browse images and performing augmentation (rotation, flipping, and brightness adjustment) to gain an additional 62,598 landmark images (details in [9]). This is a whole-image labeling data set with 8 classes: crater, dark dune, slope streak, bright dune, impact ejecta,

(a) Observation tray (b) Ground (c) Horizon (d) Mastcam cal. target

(e) Scoop (f) Turret (g) MAHLI (h) Portion tube

Figure 5.8 Examples of Mars Surface Image (Curiosity rover) data set (version 1).

swiss cheese, spider, and other (for surface features not included in any other class). Examples are shown in Fig. 5.7.

The **Mars Surface Image (Curiosity rover)** data set [10] consists of 6691 images spanning 24 classes that were collected by the Mars Science Laboratory (MSL or Curiosity) rover by three instruments (Mastcam Right eye, Mastcam Left eye, and Mars Hand Lens Imager (MAHLI)). Examples are shown in Fig. 5.8. These images are (as with the HiRISE Landmarks data set) the "browse" version of each original data product, not full resolution. They are roughly 256x256 pixels each. If desired, the full-size images can be obtained from the PDS Image Atlas at https://pds-imaging.jpl.nasa.gov/search/. The class labels can be mapped directly to the full-size products.

For reproducibility, a recommended split of the data into train, validation, and test data sets is included. The images are split according to their sol (Martian day) of acquisition, which ranges from MSL sol 3 to 1060 (August 2012 to July 2015). The division of images chronologically was employed to simulate how the system will be used operationally, with an image archive that grows over time. Wagstaff et al. [11] reported on results achieved on this data set with a convolutional neural network and additional methodology details.

The **Planetary Body Limb and Plume** data set [12] was compiled to aid in evaluating methods for the automated detection of planetary bodies and limbs (apparent edge) in images. The data set contains manually generated labels for 308 NASA images of planets and moons. The labels annotate the location of the limb (edge) of the body and any plumes emitted by the body, if any. "Plume" in this context refers to bright material emitted from the body, such as icy plumes from Enceladus or volcanic plumes from Io. Each image is labeled by the presence or lack of plumes; 112 of the 308 labeled images contain plumes. Examples are shown in Fig. 5.9.

(a) Galileo SSI observation of Jupiter's moon Callisto (b) Galileo SSI observation of Jupiter's moon Europa (c) Galileo SSI observation of Jupiter's moon Ganymede

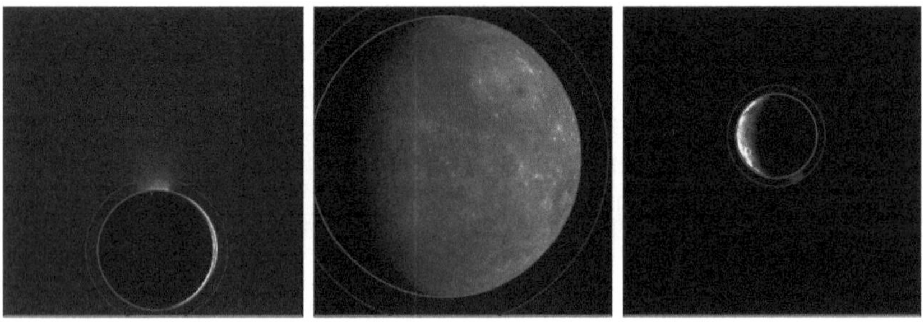

(d) Cassini ISS observationof Saturn's moon Enceladus (e) MESSENGER MDIS observation of Mercury (f) New Horizons LORRI observation of Jupiter's moon Io

Figure 5.9 Examples of the Planetary Body Limb and Plume data set. The limb of the body is shown in red (mid gray in print version) and the annulus (an extended region in which plumes may occur) is in blue (dark gray in print version). Plumes, if present, are indicated with a cyan (light gray in print version) shading inside the annulus (d, f).

5.4.2 Tools for labeling image data

There are many options for creating a new labeled image data set. Labeling a large number of images can be labor- and time-consuming, and labeling images in planetary science may require the judgment of highly skilled domain experts (e.g., to identify different types of rocks in Mars rover images or identify different parts of the Curiosity rover). Labeling a large data set usually requires more time than any single scientist can provide, so it is common to distribute the labeling effort amongst multiple people (crowd-sourcing). For example, the creators of the ImageNet data set used Amazon Mechanical Turk to distribute 1.2 million images to thousands of users for labeling into 1000 classes [13]. To ensure reliable labels, they collected at least ten labels per image and used the most commonly selected class as the final image label. When working with planetary science images, which are likely to be less familiar to nonexperts, it is important to work with scientists to define the labeling tasks and provide detailed training documentation to increase the consistency and accuracy of the labels.

There are several tools and services that provide support for obtaining crowd-sourced image labels. Here we highlight some examples that meet different project needs.

Zooniverse (https://www.zooniverse.org) is a free web platform that supports citizen science projects to crowd-source labels. Zooniverse provides customizable user interfaces for labeling data in various ways, and it also establishes a communication channel between researchers and citizen scientists. Zooniverse does not provide automated functions to accept or reject individual contributions (labels).

Amazon Mechanical Turk (AMT) (https://www.mturk.com/) is a paid web service that provides an on-demand, scalable, human workforce to complete tasks of many kinds, including image labeling. AMT provides a marketplace in which requestors set the payment for each unit of work (e.g., each image to be labeled), and workers can select which tasks they want to complete and be paid for. AMT provides tools to help select good workers, whose performance is evaluated based on previous work for other tasks.

Yandex.Toloka (https://toloka.yandex.com/) is a paid web platform for collecting and processing large amounts of data using crowd-sourcing. Yandex.Toloka is similar to AMT in that it also provides a marketplace for work. Yandex.Toloka provides comprehensive tools and customizable templates to design labeling tasks and select or filter workers by their past performance and skills.

LabelBox (https://labelbox.com/) is a company that provides solutions for labeling tasks with a focus on computer vision related problems. LabelBox provides paid support for commercial use, and it is free to use for educational institutions and noncommercial organizations. LabelBox has a graphical user interface platform with built-in and customizable

HTML/Javascript templates to support various labeling tasks. In addition, LabelBox also provides an extendable Python API to further help users create customized labeling workflows. LabelBox provides outsourced labeling capabilities by connecting labeling tasks with registered third party companies that provide labeling workforce.

Interactive Data Analyzer and Reviewer (IDAR, https://github.com/stevenlujpl/IDAR) for Machine Learning Systems is an open source tool that provides highly customizable interfaces for various tasks in the process of labeling and refining image data sets. These tools include the Class Discovery Tool, Image Labeling Tool, Ambiguous Label Analyzer, and Classification Error Analyzer. The tools make use of Python and JavaScript. Python has been proven to be efficient for data analysis tasks, and JavaScript is designed to enable interactive actions for browser-based applications. A customizable Python script and template file are used to generate a static HTML page with associated resources (Javascript, CSS, and images) to display the browser-based user interface. The static page can be shared with collaborators over email or file exchange without requiring the setup of a web server. The results of a researcher's analysis are saved in the browser using the HTML5 Storage API, and they can be exported and shared as CSV files.

5.5 Example PDS image classifier results

Once the labeled data has been obtained and its quality has been verified, we can start building machine learning models. In this section, we show an example of a deep learning model that was trained and evaluated using the publicly available Mars Orbital Image (HiRISE) data set (version 3) which was described in Section 5.4. The trained model was applied to the full HiRISE archive in early 2018 to enable the content-based image search capability for the PDS Image Atlas. It is also applied to new HiRISE images as they are delivered to the PDS. We will refer to this model as the HiRISENet model since it is trained and evaluated using HiRISE images.

5.5.1 Train, validation, and test sets

The HiRISE landmark images were divided into train, validation, and test sets while ensuring that all landmarks that were cropped from the same HiRISE source image stayed together. This reduces correlation between the train, validation, and test sets. As described above, augmentation was applied to increase the data set size. The resulting training set contains 46,970

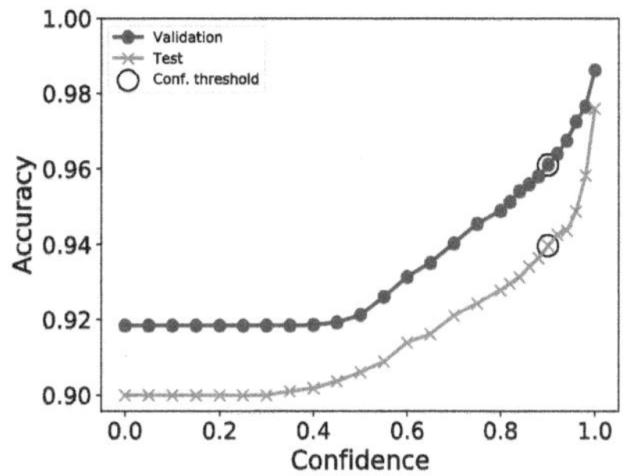

Figure 5.10 HiRISENet filtered performance: only examples with posterior probability (confidence) greater than or equal to a given threshold (x-axis) are included. The user can specify this threshold. Circles indicate performance with a confidence threshold of 0.9.

images, and validation set contains 13,391 images. Augmented images were excluded from the test set to enable comparison with results obtained prior to augmentation; it contains 1.810 images.

5.5.2 Model fine-tuning

We utilized transfer learning to adapt an existing convolution neural network to be able to classify HiRISE images. The original network, CaffeNet [14], was pretrained using 1.2 million images, spanning 1000 classes, from the ImageNet [15] database. To obtain the HiRISENet model, we trained this model for an additional 16,000 iterations on the HiRISE training and validation images to adapt ("fine-tune") it for use in this domain. The fine-tuning process involves slowing down learning for the bottom layers and focusing effort on adapting the top (output) layer to the new set of classes (here, 8 classes instead of 1000). By allowing only minimal changes (learning) to the bottom layers, we are able to leverage the basic image feature extraction (filters) that the network already learned, which generalizes to the HiRISE images. This enables us to train a model with only tens of thousands of examples instead of requiring millions. Specifically, we used a base learning rate of 0.0001, batch size 10, learning rate multiplier 1 for layer 1 to layer 7, and learning rate multiplier 10 for the final (output) layer. After exploring several values, we found that these parameter settings enabled the best generalization performance. Fig. 5.10 shows the model's performance when using the posterior probability (confidence) of each classification to filter out low-confidence predictions. The leftmost point shows performance when no examples are filtered (92% for validation and 90% for test). The circled values show performance achieved when discarding predictions with a confidence less than 0.9 (96% for validation and 94% for test).

5.5.3 Model calibration and performance

Modern convolutional neural networks are often poorly calibrated, which means that their self-reported posterior probabilities do not reliably reflect empirical probabilities [16]. To improve the reliability of posterior probabilities, we explored various model calibration methods and found that the most effective method for this data set is temperature scaling [16], which estimates a scaling parameter ("temperature") to adjust the neural network's logits prior to their conversion to probabilities. Reliability diagrams for the HiRISENet model before and after temperature scaling method are shown in Fig. 5.11. The expected calibration error (ECE) (average weighted error across the probability bins) and maximum calibration error (MCE) (maximum error across the probability bins) were improved (reduced) after temperature scaling method was applied.

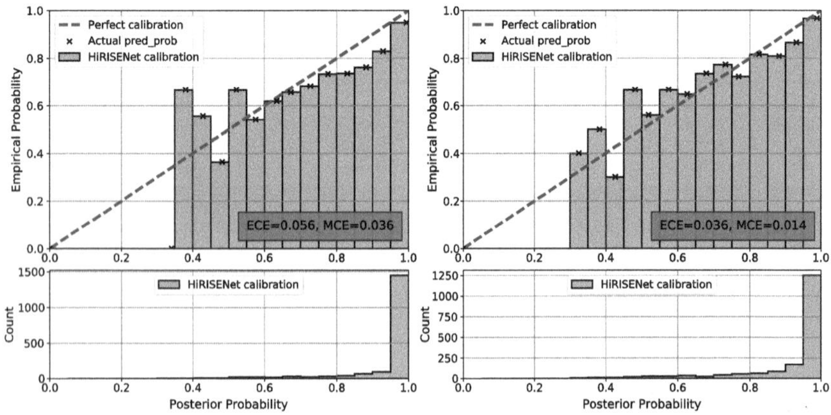

(a) Before temperature scaling (b) After temperature scaling

Figure 5.11 HiRISENet reliability diagrams.

5.5.4 Access to HiRISENet classification results

Once HiRISENet was fine-tuned and evaluated, it was used to make predictions for the entire archive of HiRISE browse images. The classification results with posterior probabilities of at least 0.9 are deployed on the PDS

Image Atlas. The following steps show an example of accessing HiRISE images that the model predicted to contain craters:

- Navigate to the PDS Image Atlas at https://pds-imaging.jpl.nasa.gov/search.
- From the "MRO HiRISE Image Content" tab, select "crater".
- Click images to the right to enlarge them. Crater bounding boxes are shown as red frame overlays.

5.6 Summary

In this tutorial, we introduced tools and services for accessing, preprocessing, and labeling PDS image data products, and we also showed an example of building a high-performance CNN image classifier using the publicly available Mars Orbital Image v3 data set. This classifier is deployed on the PDS Image Atlas to enable content-based image search.

Acknowledgments

The images in this work were obtained from the PDS Cartography and Imaging Sciences Node. This research was carried out at the Jet Propulsion Laboratory, California Institute of Technology, under a contract with the National Aeronautics and Space Administration. Copyright 2019 California Institute of Technology. U.S. Government sponsorship acknowledged.

References

[1] M.W. Libbrecht, W.S. Noble, Machine learning applications in genetics and genomics, Nature Reviews Genetics 16 (2015) 321–332.

[2] K.T. Butler, D.W. Davies, H. Cartwright, O. Isayev, A. Walsh, Machine learning for molecular and materials science, Nature 559 (2018) 547–555.

[3] K.G. Liakos, P. Busato, D. Moshou, S. Pearson, D. Bochtis, Machine learning in agriculture: a review, Sensors 18 (2018).

[4] PDS3 standards reference version 3.8, https://pds/datastandards/pds3/standards/sr/StdRef_20090227_v3.8.pdf, 2009. (Accessed 4 October 2019).

[5] R.G. Deen, The VICAR file format, https://www-mipl.jpl.nasa.gov/external/VICAR_file_fmt.pdf, 1992. (Accessed 4 October 2019).

[6] Definition of Flexible Image Transport System (FITS), https://www-mipl.jpl.nasa.gov/external/VICAR_file_fmt.pdf, 2018. (Accessed 4 October 2019).

[7] PDS4 standards reference version 1.12.0, https://pds.nasa.gov/datastandards/documents/sr/v1/StdRef_1.12.0.pdf, 2019. (Accessed 4 October 2019).

[8] The PDS4 data provider's handbook version 1.12.0, https://pds.nasa.gov/datastandards/documents/dph/v1/PDS4_DataProvidersHandbook_1.12.0.pdf, 2019. (Accessed 4 October 2019).

[9] G. Doran, S. Lu, L. Mandrake, K. Wagstaff, Mars orbital image (HiRISE) labeled data set version 3, https://doi.org/10.5281/zenodo.2538136, 2019.

[10] A. Stanboli, K. Wagstaff, Mars surface image (curiosity rover) labeled data set, https://doi.org/10.5281/zenodo.1049137, 2017.

[11] K.L. Wagstaff, Y. Lu, A. Stanboli, K. Grimes, T. Gowda, J. Padams, Deep Mars: CNN classification of Mars imagery for the PDS Imaging Atlas, in: Proceedings of the Thirtieth Annual Conference on Innovative Applications of Artificial Intelligence, 2018, pp. 7867–7872.

[12] M. Cameron, G. Doran, K. Wagstaff, Planetary body limb and plume labels for NASA images, https://doi.org/10.5281/zenodo.2556063, 2019.

[13] O. Russakovsky, J. Deng, H. Su, J. Krause, S. Satheesh, S. Ma, Z. Huang, A. Karpathy, A. Khosla, M. Bernstein, A.C. Berg, L. Fei-Fei, ImageNet large scale visual recognition challenge, International Journal of Computer Vision 115 (2015) 211–252.

[14] Y. Jia, E. Shelhamer, J. Donahue, S. Karayev, J. Long, R. Girshick, S. Guadarrama, T. Darrell, Caffe: convolutional architecture for fast feature embedding, arXiv preprint, arXiv:1408.5093, 2014.

[15] J. Deng, W. Dong, R. Socher, L.-J. Li, K. Li, L. Fei-Fei, ImageNet: a large-scale hierarchical image database, in: CVPR09, 2009.

[16] C. Guo, G. Pleiss, Y. Sun, K.Q. Weinberger, On calibration of modern neural networks, in: Proceedings of the 34th International Conference on Machine Learning, 2017, pp. 1321–1330.

CHAPTER 6

Planetary image inpainting by learning mode-specific regression models

Hiya Roy[a], Subhajit Chaudhury[b], Toshihiko Yamasaki[b], and Tatsuaki Hashimoto[a]
[a]The University of Tokyo, Department of Electrical Engineering and Information Systems, Tokyo, Japan
[b]The University of Tokyo, Department of Information and Communication Engineering, Tokyo, Japan

6.1 Introduction

In computer vision, the task of filling the missing pixels of an image is known as image inpainting. The main challenge to solve the image inpainting task is to synthesize the missing pixels in such a way that it looks visually realistic and coherent to human eyes when compared with the original ones. Image inpainting can have several applications in photo or video editing such as image-stitching [1], scene completion [2] etc. In this work, we try to solve the problem of image inpainting in the realm of Mars orbital images.

In the quest of exploring and understanding the planetary body, several missions to the Moon, Mars and other planetary bodies in the solar system, have been carried out over the years. The on-board imaging cameras, spectrometers, and other instruments from all of these missions provide us with a huge volume of planetary orbital images and other critical information that can unveil interesting information regarding the planet or the planetary bodies. With the recent advancement in imaging technology, current on-board cameras can provide very high-resolution planetary images. However, to obtain these high-resolution images, the swath width of the on-board cameras of the orbiting satellite is kept lower which in turn creates discontinuity or black missing pixel regions which we call 'artifacts' on the captured images of Moon or Mars. A schematic of one possible reason for such artifacts along with the zoomed-in and out snapshots of Lunar surface are shown in Fig. 6.1. Moreover, this kind of artifacts are not only present on the Lunar surface, but they also occur on the MRO captured

Machine Learning for Planetary Science
https://doi.org/10.1016/B978-0-12-818721-0.00015-X
Copyright © 2022 Elsevier Inc.
All rights reserved.

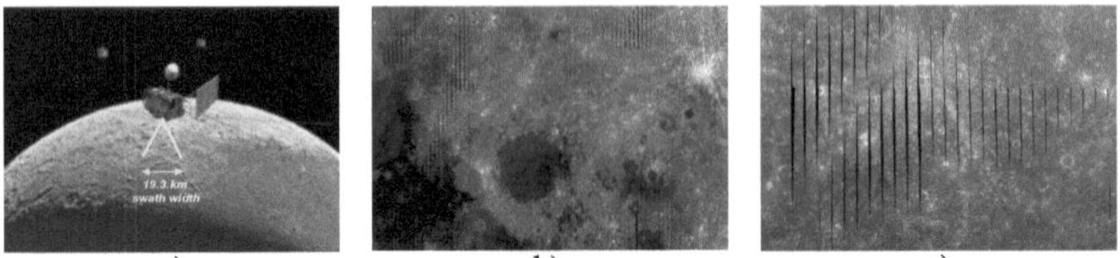

a) b) c)

Figure 6.1 a) Artist's conception for the reason behind the artifacts. Theoretically these artifacts are caused because the swath width of the on-board cameras of the orbiting satellite is kept lower to obtain the high-resolution images. The image of this on-orbit configuration of the KAGUYA main orbiter is taken from [3] b) Zoomed-in and c) zoomed-out snapshots of the artifacts on the Lunar orbital images acquired by Kaguya mission's Selene spacecraft [4].

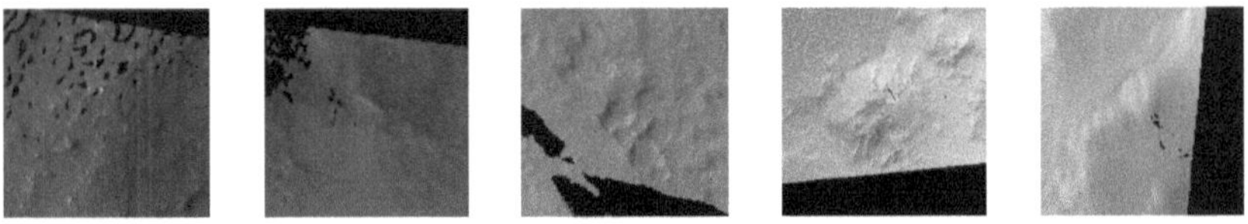

Figure 6.2 Example of artifacts on Mars orbital images.

Mars orbital image. An example of such artifacts on Mars orbital image is shown in Fig. 6.2.

In this paper, we focus on restoring the Mars orbital images that were captured by HiRISE (High-Resolution Imaging Science Experiment) camera on the Mars Reconnaissance Orbiter. We aim to intelligently predict the missing pixels in a context-aware fashion (as shown in Fig. 6.3) to get the entire orbital image. However, it should be noted that we are not recovering the real pixel values but predicting the pixel values based on the data-driven methods.

While solving this task, we faced two problems - 1) gradation (different modes) of histogram distribution in the input images, 2) missing pixels located in the extremities (any corners right/left/up/down) of the image. We tackled the first problem by clustering images with similar intensity distribution and then training regression models having expertise in restoring missing pixels in the images with that particular intensity distribution. Our intuition behind this is mode-specific encoders will provide better inpainting results when compared with only one encoder trained on an average intensity distribution [5]. The second problem lies in the location of the missing pixels which are at the extremities of an image. Due to such localization, it is difficult to perform the inpainting task because here the

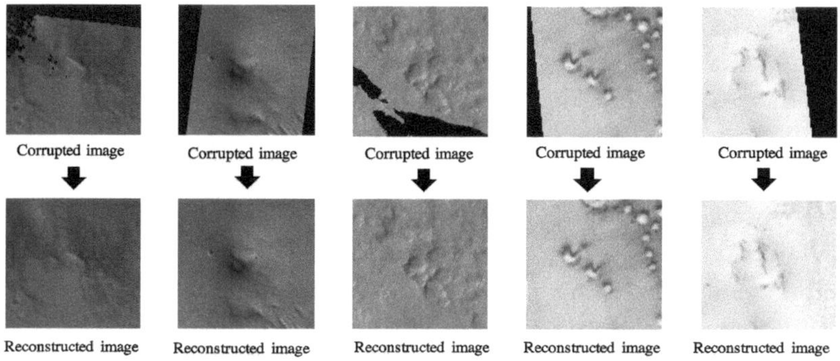

Figure 6.3 Qualitative illustration of the image inpainting task. The first and the second row shows the corrupted images and the corresponding reconstructed images respectively.

information of only one side of the missing pixel region is available, the other three sides being unavailable to the network. We solve this second problem by doing a reflection based information enhancement technique that brings the missing pixel region in the center of the image, thus artificially creating a neighborhood that has information from both sides of the missing pixel region. Experimental results showed that our proposed approach of training mode-specific expert neural networks along with the reflection-based information enhancement technique could restore images with a significant improvement in visual quality.

6.2 Related works

Humans have an extraordinary capacity to understand and visually predict the missing region of an image or scene depending on the neighboring context. Previous research works on image inpainting have shown that it is possible to learn and predict this structure of missing pixels by designing and training convolutional neural networks (CNNs) [6–9], a class of deep neural networks able to solve several vision-based tasks with outstanding performance. Xie et al. [10] proposed Stacked Sparse Denoising Autoencoders (SSDA), a combined approach of sparse coding and deep networks pretrained with denoising auto-encoder to solve the blind image inpainting task, which is a more challenging inpainting task given that the algorithm does not know the location of the missing pixels and it learns to find the location of the missing pixels and then restore them. Kohler et al. [11] showed a mask specific deep neural network-based blind inpainting technique for filling in small missing regions in an image. Chaudhury et al. [12]

attempted to solve the blind image inpainting task using a lightweight fully convolutional network (FCN) demonstrating a comparable performance with the sparse coding based K-SVD [13] technique. Pathak et al. [14] proposed Context Encoders - a channel-wise fully connected convolutional neural network-based approach, trained using both standard ℓ_2 loss pixel-wise reconstruction loss and adversarial loss [15]. This was the first method that could inpaint large holes or missing regions existing in an image by predicting missing pixels based on the context of the surrounding areas of that region. Later Iizuka et al. [16] extended the work of [14] and demonstrated that by leveraging the benefits of dilated convolution layers, a variant of standard convolutional layers network, their encoder-decoder based method could restore missing pixels that are consistent both locally and globally. Similar to [14], this approach also uses an adversarial training approach for image completion, but unlike [14], this method could handle arbitrary image size and mask because of the proposed global and local context discriminator networks. Recently, Yu et al. [17] presented a unified feedforward generative network with a novel contextual attention layer, trained with reconstruction losses and two Wasserstein GAN [18,19] and showed that the unified framework could inpaint images with multiple holes of variable sizes situated at arbitrary locations.

Meanwhile, deep image inpainting techniques have also been applied in the context of planetary surface image restoration tasks. Roy et al. [20] first proposed a U-Net based approach to restoring the missing pixels of the lunar surface image collected by the Multiband Imager (MI) instrument on-board the Kaguya satellite. In this work, we extend this idea of filling in missing pixels on the Mars orbital image by training mode-specific regression models based on the clusters of different images with similar histogram distribution on the experimental dataset.

6.3 Experimental data

In this work, we have used the Mars orbital images collected by the HiRISE camera on-board the Mars Reconnaissance Orbiter having a spatial resolution of approximately 30 cm/pixel [21]. Although this dataset has three color channels: red (550-850 nm), blue-green (400-600 nm), and near-infrared (800-1000 nm), we have used the grayscale version of the image. This dataset has a total of 73,031 landmarks, where 'landmarks' refer to different planetary surface features such as craters, bright sand dunes, dark sand dunes, dark slope streaks etc. [22]. Amongst these (73,031 landmarks),

10,433 landmarks are detected and extracted from 180 HiRISE browse images. The remaining 62,598 landmarks are the augmented version (90 degrees, 180 degrees, 270 degrees clockwise rotation, horizontal flip, vertical flip and random brightness adjusted) of 10,433 original landmarks. Each image is of size 227×227 pixels and for this experimental purpose, we resized them to size 256×256. The dataset is available at https://zenodo.org/record/2538136#.XYjEuZMzagR.

6.4 Proposed method

Here, our goal is to predict a plausible hypothesis to fill in the missing pixels of an unknown corrupted image by learning the mapping between other clean and corrupted images in the dataset. We build upon the recently proposed Context Encoder [14] and extend our method to solve an inpainting problem for a dataset having several modes in the histogram distributions. Furthermore, we propose a reflection based information enhancement methodology, which helps us to deal with the problem of the location of the missing pixel regions in the extremities of an image.

6.4.1 Unsupervised separation of histogram clusters

In our dataset, we found that the Mars images have several modes in the histogram distributions. Hence our goal was to separate the images with different intensity distribution in different clusters. The intuition behind such clustering was that regression model trained with images of particular intensity distribution such as $p_1(x)$ or $p_2(x)$ will give better performance compared to a single model with an average intensity distribution $p(x)$ as shown in Fig. 6.4a. For that purpose, first, we calculated the number of black pixels in every image and considered the images having the number of black pixels less than 5 as *clean image* and number of black pixels more than 50 as *corrupted images*. Then we performed k-means clustering [23], to cluster these images with missing pixels into different groups based on their histogram distribution as shown in the following equation:

$$h_k(x) = \frac{n(x[i,j] = k)}{\text{number of pixels}} \tag{6.1}$$

where k varies from 1 to 255 for the features and $h_0(x)$ is the number of black pixels.

To find the optimal number of clusters k, we carried out Knee point analysis [24]. As shown in Fig. 6.4b, the optimal number of image clusters

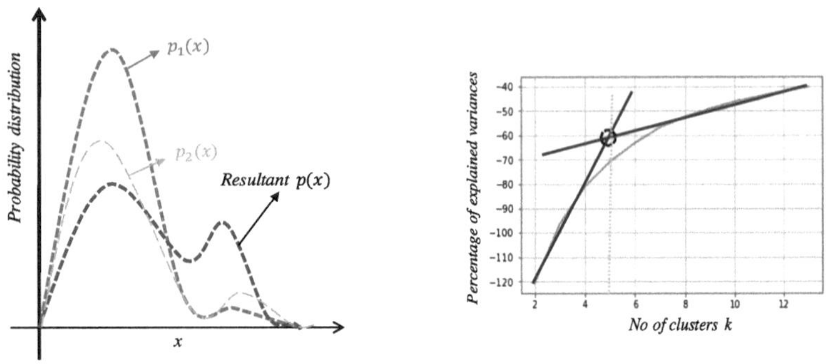

(a) Motivation behind unsupervised separation of histogram clusters.

(b) Knee point analysis.

Figure 6.4 Histogram distribution of several modes of the Mars HiRISE dataset and Knee point analysis for determining the optimal number of clusters.

Table 6.1 Number of clean and corrupted images in each cluster.

Method	Number of clean images	Number of images with missing pixels
Cluster 0	15,697	4108
Cluster 1	9733	2755
Cluster 2	3747	2060
Cluster 3	12,703	3133
Cluster 4	10,775	3598

comes out to be 5. The images corresponding to each cluster are shown in Fig. 6.5. Subsequently, from the five cluster centers in the grayscale histogram space, we assign class label to each clean image, according to the cluster center having closest Euclidean distance. The statistics of these extracted images are given in Table 6.1. Next, for each of the clusters, we extracted a mask of the missing pixels from the corrupted images and randomly superimposed them on clean images and created pairs of clean and corrupted images for training and testing for five different regression models. An example of such artificially created corrupted image is shown in Fig. 6.6.

6.5 Network architecture

For solving the inpainting task of planetary images having different intensity distribution, we train generative adversarial network (GAN) based

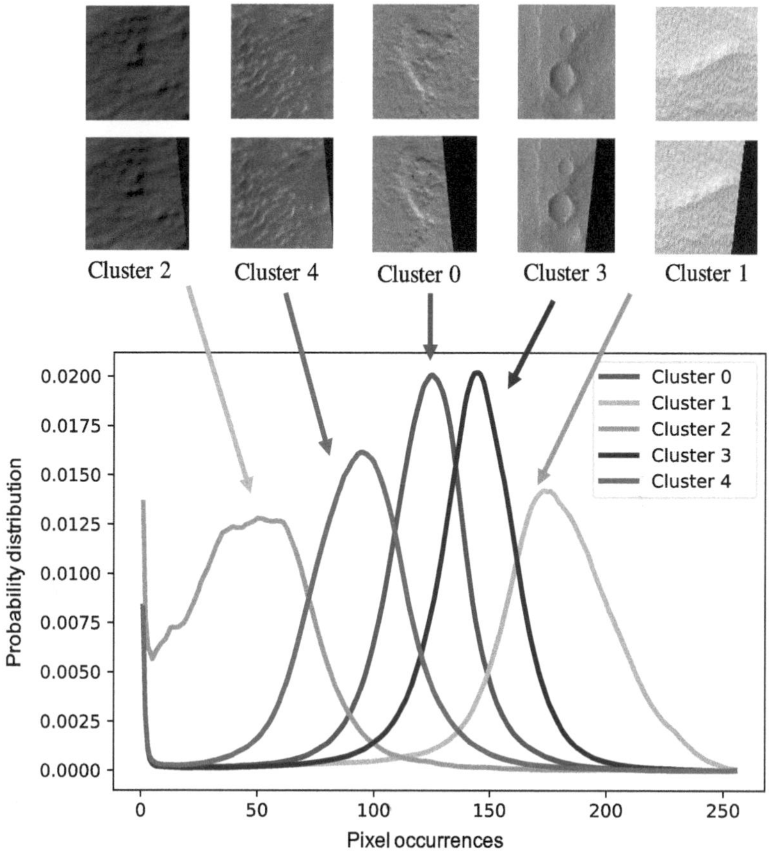

Figure 6.5 Histogram distribution of different images in the experimental data set and the example of clean and corrupted images belonging to each of the five clusters divided based on their intensity distribution.

models [15] which has shown promising results in generative modeling of images [25] in recent years. The overview of our proposed method is shown is Fig. 6.7. For the generator network, we adopt U-Net architecture from [26] which consists of a contraction and an expansion path. Here, the contraction path consists of a sequence of two 3×3 convolution operation, each followed by a nonlinear activation (ReLU) and a 2×2 max-pooling operation with stride 2. The expansion path consists of a sequence of 2×2 upconvolutions and concatenations with correspondingly cropped feature maps from the contracting path and two 3×3 convolutions, each followed by ReLU layer [26]. In total, our generator network has 28 convolutional

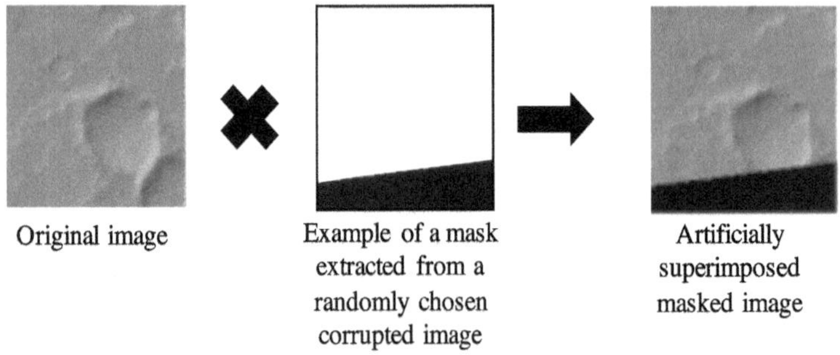

Figure 6.6 Example of artificially created corrupted/masked images.

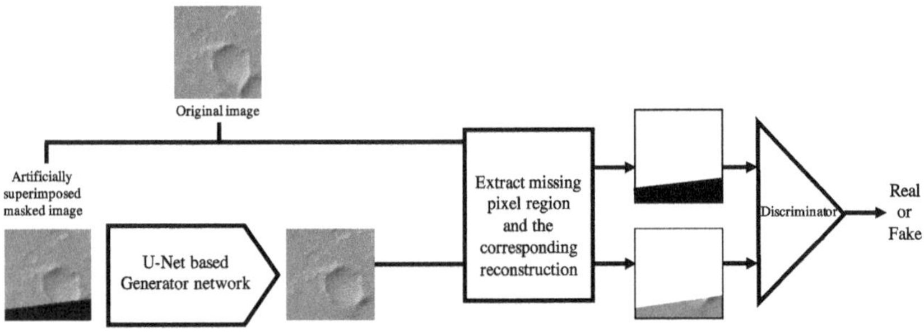

Figure 6.7 Overview of the proposed method. Our model has a generator and a discriminator network and is built upon the idea of Context Encoder [14] for planetary image inpainting task.

layers. For the discriminator network, we use three convolution layers, each followed by a LeakyReLU activation function and a dense layer.

6.5.1 Training details

We trained our model using adversarial loss [15] to learn the mapping between input corrupted image X and output clean image Y. Let us denote the input and output data distribution as $x \sim p_{data}(x)$ and $y \sim p_{data}(y)$. To learn the mapping between $G: X \rightarrow Y$ and the discriminator D_Y, we solve the following objective function:

$$\min_{G} \max_{D_Y} \text{Ł}_{GAN}(G, D_Y, X, Y) = \mathbb{E}_{y \sim p_{data}(y)}[\log D_Y(y)] + \mathbb{E}_{x \sim p_{data}(x)}[\log(1 - D_Y(G(x))] \tag{6.2}$$

where the generator network tries to generate images $G(x)$ looking similar to the output clean images from the distribution Y and the discriminator

D_Y network tries to differentiate if the generated image is real or fake. Here "real" refers to the original clean/uncorrupted image and "fake" refers to the artificially restored image. While the generator tries to minimize the objective function, the discriminator tries to maximize it, thus playing a min-max game. We used five different models for five different clusters (which we got from Knee point analysis as described in Section 6.4.1) and trained them for 30,000 epochs by regressing the ground truth content of the missing pixel region. For optimization, we used the recently proposed stochastic gradient descent solver, Adam optimizer [27] with a batch size of 8 and a learning rate of 0.0002. We implement our models using Tensorflow and Keras library in Python.[1] Both the generator and the discriminator networks were trained together on a GTX-1080 GPU for approximately 8 hours.

Initially, we trained the models using images of size 256×256. However, one major challenge in solving planetary image inpainting is the location of the missing pixels which is situated at the extremities of the image as shown in the first row of Fig. 6.3. This causes the unavailability of information from all sides of the missing pixel region. Since the state-of-the-art inpainting algorithms in the literature mostly focus on solving inpainting in the center region, they fail to restore the missing pixels located at the extremities of the image. To solve this problem, we propose a reflection based information enhancement methodology.

6.5.2 Reflection based information enhancement methodology

By using the reflection-based information enhancement methodology, we augment the corrupted images in such a way that the missing pixel region comes in the center portion of the image. We follow the set of rules as shown in Fig. 6.8. After applying this technique, we retrain the proposed networks using a pair of clean and corrupted images of size 256×512 for all of the five clusters. This method helps the network to get information from both sides, thus significantly improving the visual quality as well as the restoration performance as shown in Section 6.6.

[1] The code, trained models and more inpainting results are available at https://github.com/hiyaroy12/planetary-image-inpainting.

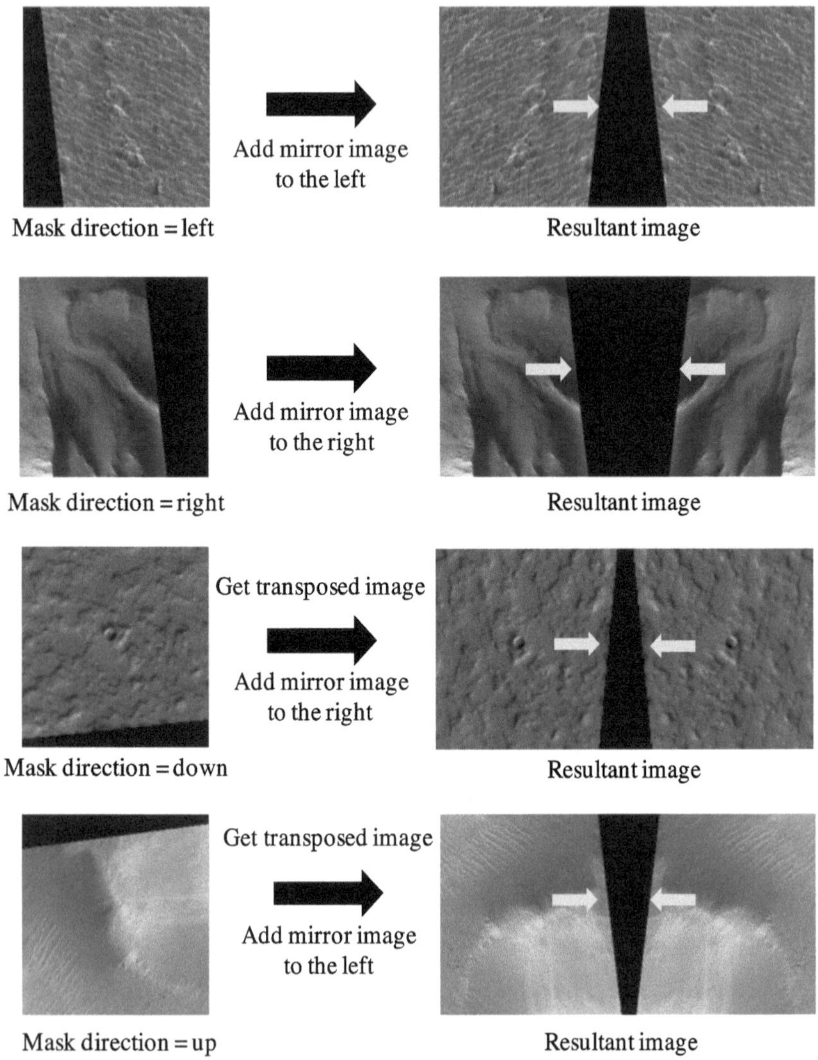

Figure 6.8 Rules for corrupted image augmentation. This helps in the information flow from both sides making it easier for the network to learn from the neighborhood and predict missing pixels in coherence with its surroundings.

6.6 Experimental results

To verify the effectiveness of our proposed algorithm we perform both qualitative and quantitative evaluation. We calculate the PSNR values of the restored images compared to their original uncorrupted images for all

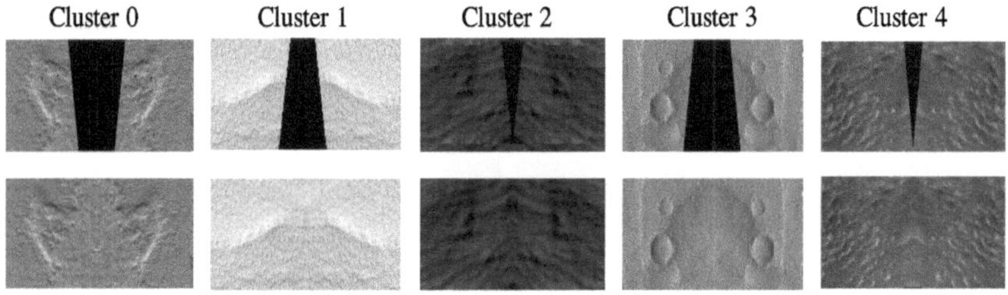

Figure 6.9 Example of resultant images for all the clusters. Row 1 and 2 show the resultant corrupted and clean images of size 256×512 respectively.

of the five clusters. Fig. 6.9 shows the example of the resultant reconstructed images of size 256×512 for each cluster, where the first row shows the resultant corrupted images obtained from the process as explained in Section 6.5.2 and the second row shows the reconstructed images obtained from the proposed algorithm. Since the original images are of size 256×256, we perform a reverse operation of the rules explained in 6.5.2 to get back the restored images of size 256×256 corresponding to their original 256×256 images.

6.6.1 Performance evaluation

We report the quantitative performance of our method in terms of peak-signal-to-noise ratio (PSNR) and structural similarity (SSIM) index. The peak-signal-to-noise ratio (PSNR) is measured in terms of Mean squared error (MSE) and is still the most common quality measure for reconstructed images. PSNR of a reconstructed image is given by:

$$PSNR = 20\log_{10}\left(\frac{MAX_I}{\sqrt{MSE}}\right) \tag{6.3}$$

where MAX_I the maximum value of pixel in the original image. A higher PSNR normally indicates higher quality reconstruction. We also calculate the structural similarity (SSIM) index [28] which provides the quantitative assessment on the perceptual quality of the reconstructed image.

Fig. 6.10 - 6.14 demonstrate the qualitative results from each cluster, where the first row represents the original uncorrupted images, the second row depicts the artificially generated corrupted images along with their PSNR values, and the third row shows the restored images with their corresponding PSNR values. These visualizations prove the ability of the context encoder models to fill in semantic details of images with missing

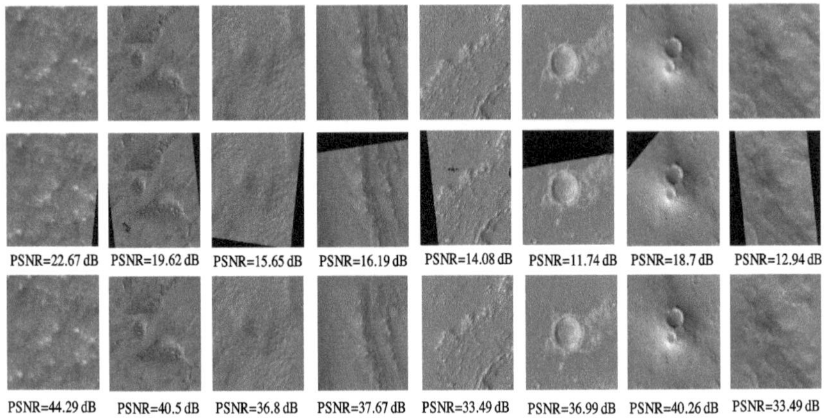

Figure 6.10 Original, corrupted and reconstructed images in Cluster 0.

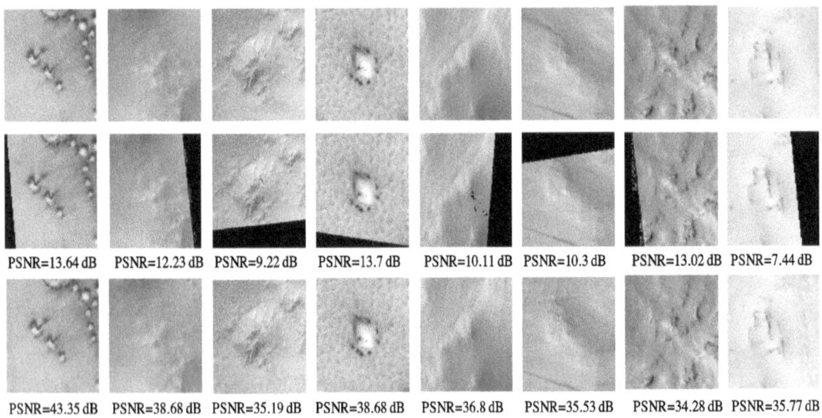

Figure 6.11 Original, corrupted and reconstructed images in Cluster 1.

regions in coherence with the surrounding pixel values. Table 6.2 and 6.3 illustrate the mean PSNR and mean SSIM values for the corrupted images and their corresponding reconstructions when trained and tested on a different set of cluster combinations. Here, at par with our expectation, the reconstructions for all the five clusters have better PSNR values compared to their corresponding corrupted images. Moreover, we find that for most of the cases, the best mean PSNR values for the restored images can be achieved when trained and tested on the same image clusters having the same intensity distribution.

Table 6.2 Mean PSNR values for the corrupted and restored images of different clusters. The best PSNR values for each cluster are in bold.

Trained on	Tested on ->	Cluster 0	Cluster 1	Cluster 2	Cluster 3	Cluster 4
Cluster 0	Mean PSNR_ corrupted_img	16.23	11.51	21.53	14.90	17.95
	Mean PSNR_ reconstructed_img	**37.53**	32.22	28.67	36.23	33.32
Cluster 1	Mean PSNR_ corrupted_img	16.23	11.51	21.53	14.90	17.95
	Mean PSNR_ reconstructed_img	35.21	**36.06**	24.54	36.03	29.43
Cluster 2	Mean PSNR_ corrupted_img	16.23	11.51	21.53	14.90	17.95
	Mean PSNR_ reconstructed_img	33.59	24.52	**36.64**	31.36	32.79
Cluster 3	Mean PSNR_ corrupted_img	16.23	11.51	21.53	14.90	17.95
	Mean PSNR_ reconstructed_img	32.79	33.20	30.64	**37.14**	33.17
Cluster 4	Mean PSNR_ corrupted_img	16.23	11.51	21.53	14.90	17.95
	Mean PSNR_ reconstructed_img	34.41	23.16	**34.43**	30.17	34.11

Table 6.3 Mean SSIM values for the corrupted and restored images of different clusters.

Trained on	Tested on ->	Cluster 0	Cluster 1	Cluster 2	Cluster 3	Cluster 4
Cluster 0	Mean SSIM_ corrupted_img	0.80	0.80	0.81	0.83	0.81
	Mean SSIM_ reconstructed_img	0.98	0.98	0.96	0.99	0.98
Cluster 1	Mean SSIM_ corrupted_img	0.80	0.80	0.81	0.83	0.81
	Mean SSIM_ reconstructed_img	0.98	0.98	0.95	0.97	0.96
Cluster 2	Mean SSIM_ corrupted_img	0.80	0.80	0.81	0.83	0.81
	Mean SSIM_ reconstructed_img	0.97	0.97	0.97	0.96	0.97
Cluster 3	Mean SSIM_ corrupted_img	0.80	0.80	0.81	0.83	0.81
	Mean SSIM_ reconstructed_img	0.95	0.98	0.96	0.98	0.98
Cluster 4	Mean SSIM_ corrupted_img	0.80	0.80	0.81	0.83	0.81
	Mean SSIM_ reconstructed_img	0.97	0.96	0.97	0.98	0.98

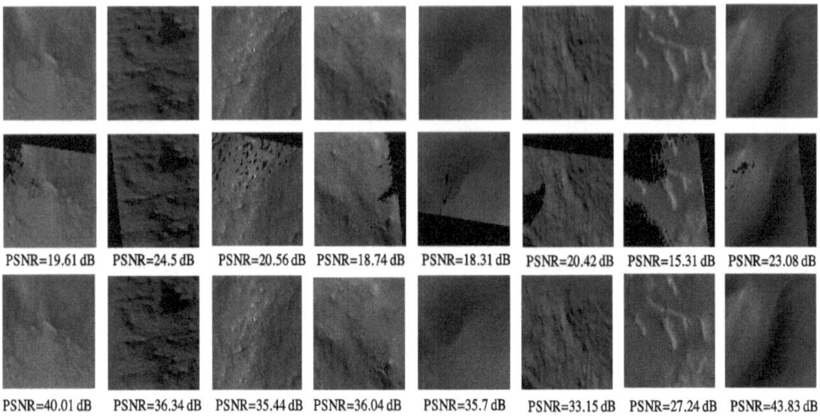

Figure 6.12 Original, corrupted and reconstructed images in Cluster 2.

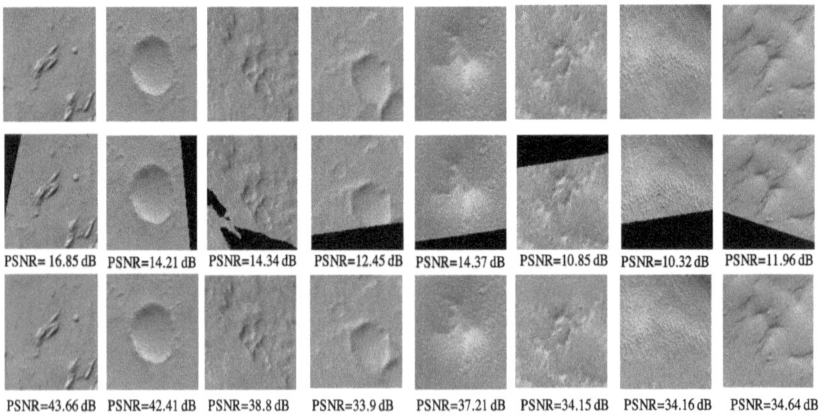

Figure 6.13 Original, corrupted and reconstructed images in Cluster 3.

Table 6.4 Comparison of reconstruction quality (PSNR values) of our method with the U-Net model [26]. Our method restores images with better visual quality compared to U-Net.

Mean PSNR values of Corrupted and Restored images										
	Cluster 0		Cluster 1		Cluster 2		Cluster 3		Cluster 4	
	Corr img	Rest img	Corr img	Rest img	Corr img	Rest img	Corr img	Rest img	Corr img	Rest img
Ours	16.23	**37.53**	11.51	**36.06**	21.53	**36.64**	14.9	**37.14**	17.95	**34.11**
U-Net [26]	16.63	20.43	12.0	17.69	21.82	17.89	14.04	19.84	18.50	20.72

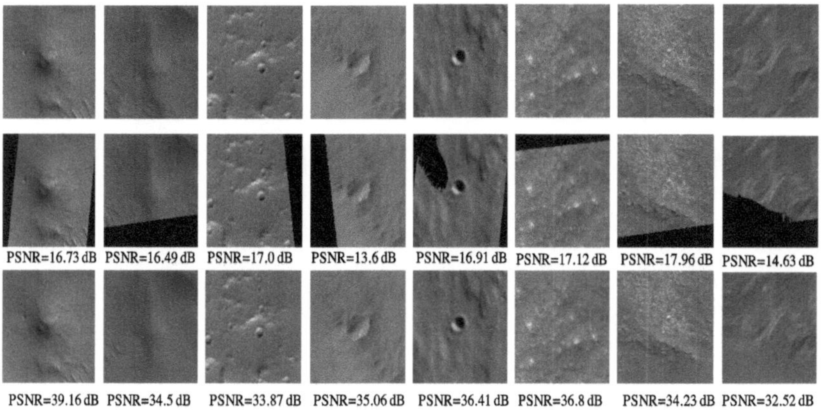

Figure 6.14 Original, corrupted and reconstructed images in Cluster 4.

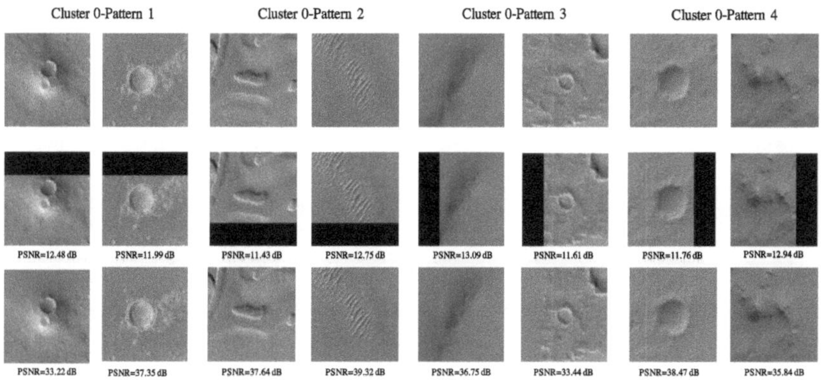

Figure 6.15 Original, corrupted and reconstructed images in Cluster 0 with four different masks.

We also compare our method with [26] and show the results in Table 6.4. We see that our method can successfully restore missing pixels at the extremities with good PSNR values in all the five cases outperforming the standard ℓ_2 loss based [26] method. We think the reason behind the better performance of our method is that it uses adversarial loss which tries to make prediction look real in contrary to standard ℓ_2 loss that tries to capture the overall structure of the missing pixel region in coherence with its neighborhood context and then tends ends to average out the multiple modes in predictions. Moreover, our method takes advantage of the proposed reflection-based information enhancement methodology which

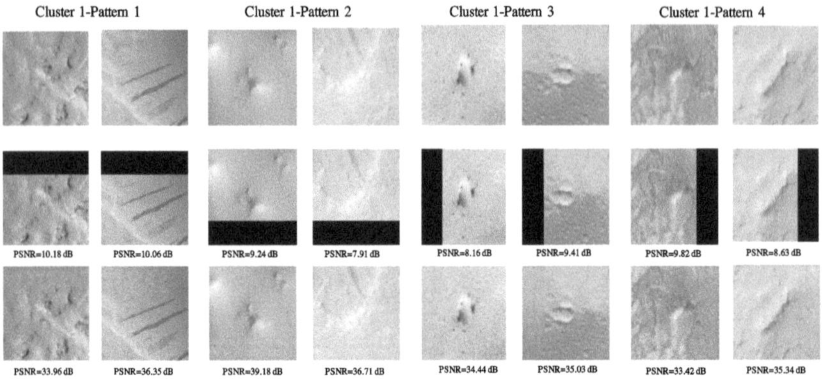

Figure 6.16 Original, corrupted and reconstructed images in Cluster 1 with four different masks.

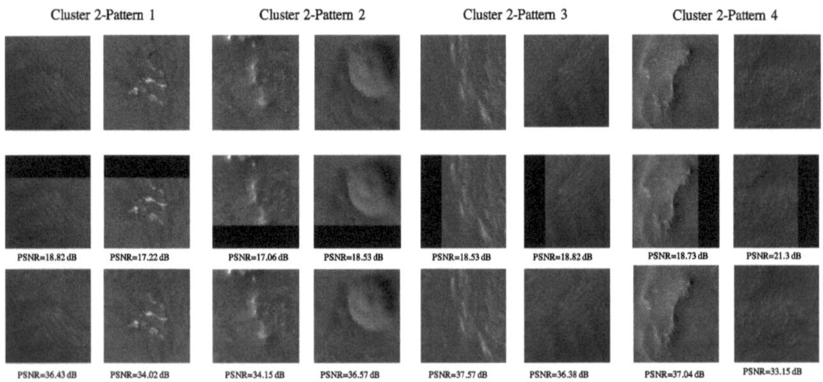

Figure 6.17 Original, corrupted and reconstructed images in Cluster 2 with four different masks.

helps in better information flow from both sides of the missing pixel thus improving the performance of our network in case of inpainting the missing pixel region located at the extremity of an image. To confirm the effectiveness of our model, we tested using four different types of masks that were never seen in the train data. We use four different types of unseen patterns consisting of 25% masking on left, right, top and bottom as shown by the examples in the second row of Fig. 6.15 - 6.19. The first and third row of Fig. 6.15 - 6.19 depict the original, and reconstructed images along with the corresponding PSNR values. We can see that our models can also restore these unseen patterns with good PSNR values.

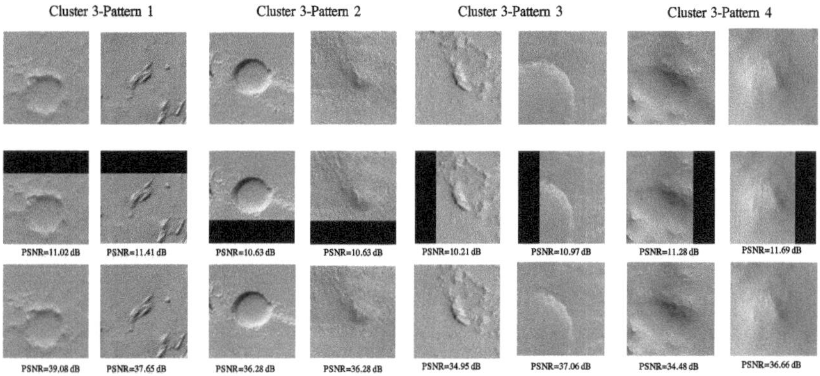

Figure 6.18 Original, corrupted and reconstructed images in Cluster 3 with four different masks.

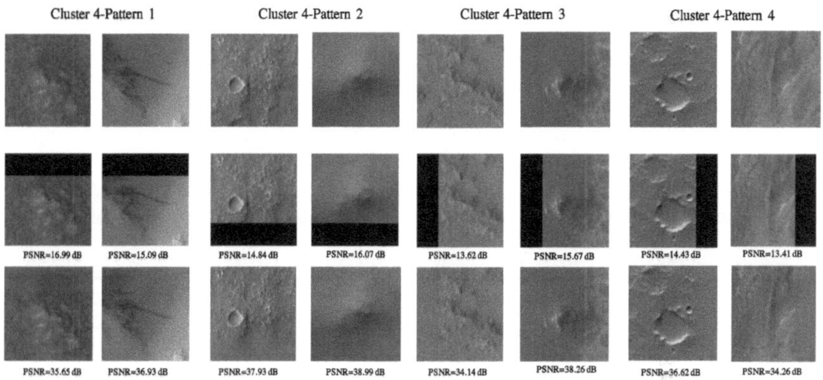

Figure 6.19 Original, corrupted and reconstructed images in Cluster 4 with four different masks.

6.7 Conclusion

In this work, we show that the missing pixels of the Mars orbital images acquired using the HiRISE camera on-board MRO can be predicted and restored successfully using the GAN based deep convolutional neural networks. Experimental results demonstrate that our proposed method can solve the inpainting task for images from various intensity distribution. We also depict that the reflection-based information enhancement technique makes it easier for the network to predict the missing pixels located at the extremities of the input image by providing extra information from the neighborhood and thus assists in filling in the missing pixel regions with impressive visual and perceptual quality. We believe the application of such image inpainting techniques will benefit the planetary science community to analyze the planetary surfaces in a more precise way.

References

[1] A. Levin, A. Zomet, S. Peleg, Y. Weiss, Seamless image stitching in the gradient domain, in: European Conference on Computer Vision, Springer, 2004, pp. 377–389.
[2] J. Hays, A.A. Efros, Scene completion using millions of photographs, ACM Transactions on Graphics (TOG) 26 (3) (2007) 4.
[3] S.-I. Sobue, S. Sasaki, M. Kato, H. Maejima, H. Minamino, H. Konishi, H. Otake, S. Nakazawa, N. Tateno, H. Hoshino, et al., The result of SELENE (Kaguya) development and operation, Recent Patents on Space Technology 1 (2009) 12–22.
[4] Y. Takizawa, Kaguya (SELENE) mission overview, in: Proceedings of the 26th International Symposium on Space Technology and Science (ISTS), June 2008, 2008.
[5] S. Roth, M.J. Black, Fields of experts: a framework for learning image priors, in: 2005 IEEE Computer Society Conference on Computer Vision and Pattern Recognition (CVPR'05), vol. 2, Citeseer, 2005, pp. 860–867.
[6] A. Krizhevsky, I. Sutskever, G.E. Hinton, Imagenet classification with deep convolutional neural networks, in: Advances in Neural Information Processing Systems, 2012, pp. 1097–1105.
[7] K. Simonyan, A. Zisserman, Very deep convolutional networks for large-scale image recognition, arXiv preprint, arXiv:1409.1556, 2014.
[8] C. Szegedy, W. Liu, Y. Jia, P. Sermanet, S. Reed, D. Anguelov, D. Erhan, V. Vanhoucke, A. Rabinovich, Going deeper with convolutions, in: Proceedings of the IEEE Conference on Computer Vision and Pattern Recognition, 2015, pp. 1–9.
[9] K. He, X. Zhang, S. Ren, J. Sun, Deep residual learning for image recognition, in: Proceedings of the IEEE Conference on Computer Vision and Pattern Recognition, 2016, pp. 770–778.
[10] J. Xie, L. Xu, E. Chen, Image denoising and inpainting with deep neural networks, in: Advances in Neural Information Processing Systems, 2012, pp. 341–349.
[11] R. Köhler, C. Schuler, B. Schölkopf, S. Harmeling, Mask-specific inpainting with deep neural networks, in: German Conference on Pattern Recognition, Springer, 2014, pp. 523–534.
[12] S. Chaudhury, H. Roy, Can fully convolutional networks perform well for general image restoration problems?, in: 2017 Fifteenth IAPR International Conference on Machine Vision Applications (MVA), IEEE, 2017, pp. 254–257.
[13] J. Mairal, M. Elad, G. Sapiro, Sparse representation for color image restoration, IEEE Transactions on Image Processing 17 (1) (2007) 53–69.
[14] D. Pathak, P. Krahenbuhl, J. Donahue, T. Darrell, A.A. Efros, Context encoders: feature learning by inpainting, in: Proceedings of the IEEE Conference on Computer Vision and Pattern Recognition, 2016, pp. 2536–2544.
[15] I. Goodfellow, J. Pouget-Abadie, M. Mirza, B. Xu, D. Warde-Farley, S. Ozair, A. Courville, Y. Bengio, Generative adversarial nets, in: Advances in Neural Information Processing Systems, 2014, pp. 2672–2680.
[16] S. Iizuka, E. Simo-Serra, H. Ishikawa, Globally and locally consistent image completion, ACM Transactions on Graphics (ToG) 36 (4) (2017) 107.
[17] J. Yu, Z. Lin, J. Yang, X. Shen, X. Lu, T.S. Huang, Generative image inpainting with contextual attention, in: Proceedings of the IEEE Conference on Computer Vision and Pattern Recognition, 2018, pp. 5505–5514.

[18] M. Arjovsky, S. Chintala, L. Bottou, Wasserstein GAN, arXiv preprint, arXiv:1701.07875, 2017.
[19] I. Gulrajani, F. Ahmed, M. Arjovsky, V. Dumoulin, A.C. Courville, Improved training of Wasserstein gans, in: Advances in Neural Information Processing Systems, 2017, pp. 5767–5777.
[20] H. Roy, S. Chaudhury, T. Yamasaki, D. DeLatte, M. Ohtake, T. Hashimoto, Lunar surface image restoration using u-net based deep neural networks, arXiv preprint, arXiv:1904.06683, 2019.
[21] A.S. McEwen, E.M. Eliason, J.W. Bergstrom, N.T. Bridges, C.J. Hansen, W.A. Delamere, J.A. Grant, V.C. Gulick, K.E. Herkenhoff, L. Keszthelyi, et al., Mars reconnaissance orbiter's high resolution imaging science experiment (hirise), Journal of Geophysical Research: Planets 112 (E5) (2007).
[22] K.L. Wagstaff, Y. Lu, A. Stanboli, K. Grimes, T. Gowda, J. Padams, Deep Mars: Cnn classification of Mars imagery for the pds imaging atlas, in: Thirty-Second AAAI Conference on Artificial Intelligence, 2018.
[23] J.A. Hartigan, M.A. Wong, Algorithm as 136: a k-means clustering algorithm, Journal of the Royal Statistical Society. Series C (Applied Statistics) 28 (1) (1979) 100–108.
[24] Q. Zhao, V. Hautamaki, P. Fränti, Knee point detection in BIC for detecting the number of clusters, in: International Conference on Advanced Concepts for Intelligent Vision Systems, Springer, 2008, pp. 664–673.
[25] A. Radford, L. Metz, S. Chintala, Unsupervised representation learning with deep convolutional generative adversarial networks, arXiv preprint, arXiv:1511.06434, 2015.
[26] O. Ronneberger, P. Fischer, T. Brox, U-net: convolutional networks for biomedical image segmentation, in: International Conference on Medical Image Computing and Computer-Assisted Intervention, Springer, 2015, pp. 234–241.
[27] D.P. Kingma, J. Ba, Adam: a method for stochastic optimization, arXiv preprint, arXiv:1412.6980, 2014.
[28] Z. Wang, A.C. Bovik, H.R. Sheikh, E.P. Simoncelli, et al., Image quality assessment: from error visibility to structural similarity, IEEE Transactions on Image Processing 13 (4) (2004) 600–612.

CHAPTER 7

Automated surface mapping via unsupervised learning and classification of Mercury Visible–Near-Infrared reflectance spectra

Mario D'Amore[a] and Sebastiano Padovan[a,b,c]
[a]German Aerospace Center (DLR), Berlin, Germany
[b]EUMETSAT, Darmstadt, Germany
[c]WGS, Darmstadt, Germany

7.1 Introduction

The sheer amount of data returned by scientific missions aimed at exploring the solar system and observing exoplanets in recent decades overwhelms classical methods to explore and discover important scientific aspects of the target body. As an example, the Mercury data return for Mariner 10 was less than 100 MB, while MESSENGER delivered about 23 TB. Future missions are expected to exceed this limit. In addition, there is a trend of increasing complexity in the data itself, e.g., going from of the Mariner 10 to the hyperspectral datasets expected from BepiColombo. This situation clearly indicates that some form of automated analysis would be beneficial, provided it is able to save time without a loss of the information content of the data.

Keeping the focus on hyperspectral remote sensing data, the typical approach for analyzing this kind of data is to model the observed radiation with a forward radiative model (like Hapke, as in [13]) or attempt to reproduce the observed radiation by setting up relevant samples in a laboratory setting using chemical and/or geomorphological context information (e.g., [15]). Complex forward models that are able to take into account the relevant physics are typically computationally intensive and difficult to use to investigate the very large parameter space covered by hyperspectral data. This consideration is even more relevant for laboratory investigations: physical simulation needs the target to be physically fabricated, hence more and

Machine Learning for Planetary Science
https://doi.org/10.1016/B978-0-12-818721-0.00016-1
Copyright © 2022 Elsevier Inc.
All rights reserved.

more parameters mean more experiments and more time. Models need computational power to be calculated in a reasonable amount of time, but could be distributed on several machines to overcome this limitation. This workaround is not effective for laboratory experiment, because most only few places meets of the environment needed for space sample simulation, like high-vacuum, -temperature, -radiation and so on.

Without a way to efficiently and rapidly explore large amounts of complex data, it is likely that valuable information will be missed in large hyperspectral data sets.

Geological maps are the gold standard for remote planetary surface studies, but producing them is an extremely time-consuming task. This process can suffer from user bias and typically only uses a few data points (e.g., 3-channel images) to describe different units. For example, Denevi et al. [6] mapped the distribution and extent of major terrain types of Mercury using MESSENGER Mercury Dual Imaging System (MDIS) camera observations of Mercury. While the camera has 11 spectral bands, the maps typically used for the terrain differentiation are RGB, where 3 representative spectral bands are mapped onto the three image color channels.

Geomorphological maps take in account additional features like surface roughness and crater density as a proxy for the age, where the correlation between age and crater density is derived from models (e.g., [3,17]). Automated techniques are becoming more common in planetary science applications, as this book testifies, and the aim of this chapter is to illustrate how to apply unsupervised learning techniques to remote sensing data. This approach requires minimal user interaction and yields scientifically interesting products like classification maps that can be directly compared with geomorphological maps and models. We present an analysis of spectral reflectance data of Mercury's surface collected by the Mercury Atmospheric and Surface Composition Spectrometer (MASCS) instrument during orbital observations of the NASA MESSENGER mission between 2011 and 2015 [20]. MASCS is a three sensor point spectrometer with a spectral coverage from 200 nm to 1450 nm. After a brief overview of the instrument and its significance for the investigation of Mercury (Section 7.2), we will illustrate how we extract and resample the data to a format useful for our ML application (Section 7.3). Then we show how to compress the data (Section 7.4.1), how to project them to a lower number of dimensions (Section 7.4.2), and finally, how to group "similar" data points together to discover salient spectral classes and their distribution on the surface. We conclude in Section 7.4.4 by providing a basic comparison of the result of the discovered spectral class distribution with maps of the surface of Mercury obtained using classical methods, in order to provide a first assessment of the machine learning techniques presented here.

7.2 Mercury and the MASCS instrument

Surface mineralogy and composition are important indicators of the past of a planetary body, since they provide hints about the processes that formed and altered the crust, which is largely the result of the interior evolution. For example, the possibility of identifying specific mineral assemblage like metamorphic rocks, which are known to form in specific pressure and temperature conditions, would provide indications on the physical processes occurring in the subsurface that produced those rocks and later transported the rocks to the surface (e.g., [24]). Similarly, observations of hydrated minerals can be interpreted as indicating the possible past presence of water, as in the case of Mars [23].

While some investigations have been published on Mercury's surface mineralogy (e.g., [12,24,33,37]), its link to the endogenous (e.g., mantle convection) and exogenous (e.g., impacts) processes that operated during the history of the planet is still difficult to elucidate (e.g., [29]). A relevant example is the geological features known as hollows, discovered on the surface of Mercury in MESSENGER data. Hollows are rimless depressions with flat floors, surrounded by halos of high-albedo material, and typically found in clusters [4]. Given this evidence, their formation mechanism likely includes the loss of volatile material through one or more processes such as sublimation, space weathering, outgassing, or pyroclastic flow. Hollows are associated with a particular spectral signature in MESSENGER's MDIS camera [37], but a specific spectral signature in spectrometer data could not be identified due to the coarse spatial resolution of the spectrometer. Overall, the only clear inference based on VNIR spectra obtained by the MASCS instrument is that Mercury's surface shows little variation, displaying no distinct spectral features except for the possible indication of sulfide mineralogy within the hollows [37].

MASCS consists of a small Cassegrain telescope with an effective focal length of 257 mm and a 50-mm aperture that simultaneously feeds an UltraViolet and Visible Spectrometer (UVVS) and a Visible and InfraRed Spectrograph (VIRS) channel. VIRS is a fixed concave grating spectrograph with a focal length of 210 mm, equipped with a beam splitter that simultaneously disperses the light onto a 512-element sensor (VIS,

300–1050 nm) and a 256-element infrared sensor array (NIR, 850–1450 nm). Data obtained by MASCS covers almost the entire surface of Mercury. The spatial resolution is highly latitude dependent due to the very elliptical orbit of the spacecraft, but a reference value ~ 5 km. This low spatial resolution is a trade-off for higher spectral resolution and more spectral channels compared to the imaging instruments (i.e., the MDIS).

The NIR sensor is characterized by $3 - 5$ times lower signal-to-noise ratios (SNRs) than the VIS detector and does not add significant information to the VIS sensor in our tests. NIR measurement can be linked and corrected to match corresponding VIS measurements following. (However, see [2], for a successful VIS/NIR cross correction.) The biggest obstacle is that the most accurate photometric corrections are only available for the VIS channel (see [9,10]). We then analyzed only data from VIS channel, that is enough for the sake of illustrating unsupervised learning techniques.

7.3 Data preparation

In what follows, we use "feature" to refer to a single spectral reflectance value in a given spectrum and refer to the spectrum as an "observation". An observation is taken on a particular area of the surface of Mercury at a given time and thus with a given illumination condition. MASCS data are obtained as a collection of binary files following the PDS3 standard [25]. We need to preprocess these to be represented as a matrix of shape $[N \times M]$ where M is the number of features and N is the number of observations before they can be used as input to typical machine learning algorithms. A spectrum can be represented as a vector, with the single spectral channels being its components, i.e., $\hat{x} = (x_1, ..., x_{512})$, for the case where the spectrum $\hat{x}$ has 512 components. The entire MASCS dataset comprises about 5 million spectra, each with 512 channels. The coverage is not homogeneous across Mercury's surface, as illustrated in Fig. 7.1a. In this section we describe how to create a homogeneous gridded representation of the observations across the entire surface so that we can apply our clustering algorithm to the entire surface, as shown in Fig. 7.1b.

A typical workflow for doing a planetary science analysis starts with the definition of a number of Regions Of Interests (ROIs), followed by a search for all data points that fall within these ROIs. This approach facilitates the extraction of spectral signatures that are specific to user-defined geological units in order to explore their properties in the context of different ROIs. For this chapter, we developed a workflow that starts by extracting the

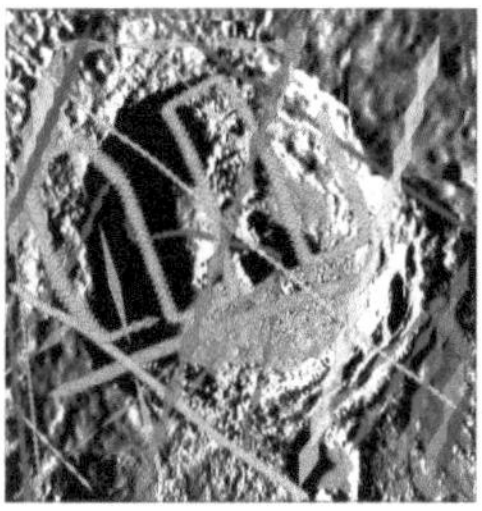

(a)

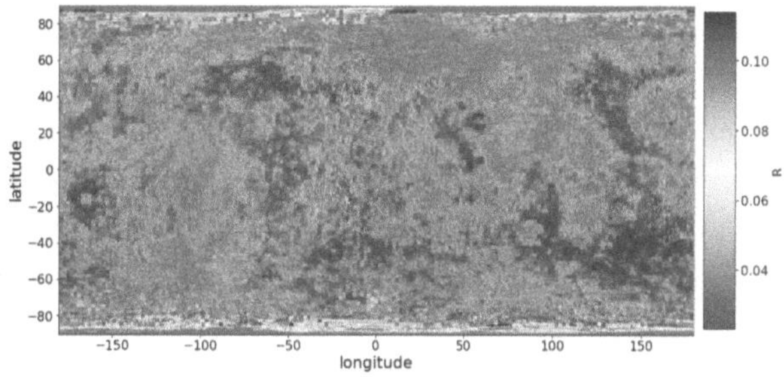

(b)

Figure 7.1 a. Example of MASCS VIRS observations in the region of the Kuiper crater, a 62-km-diameter crater with a central peak located at 11.35°S 31.23°W on Mercury. Kuiper is the brightest landmark on the planet and has a fresh ray system, suggesting that it is among the youngest craters of Mercury [8,26]. Each polygon approximately represents a VIS and NIR observation (in reality the two channel are slightly misaligned). Different colors represent different spacecraft orbits. Note how different orbits show different polygon sizes, probably due to different observation geometry or spacecraft altitude. **b.** Square polygons are used as base pixels to construct a global datacube, using the median value of all the observations within each square as the final pixel reflectance. This plot shows reflectance values from the 700 nm band of VIS.

data from the original files using the GDAL[1] python binding. We then organized them in a PostgreSQL relational database,[2] with the PostGIS spatial extension.[3] The database is currently hosted at the DLR in Berlin.

First, the whole dataset of ~ 5 million spectra is resampled to a planet-wide rectangular grid of $1° \times 1°$ in the latitudinal band between $\pm 80°$. The cell longitudinal size varies between ~ 40 km at the equator to a minimum of ~ 10 km at $\pm 80°$. Thus, the area spanned by each grid cell depends on the latitude. However, the same is true for the acquisition process, where higher spatial resolution is reached near the equator and lower resolution at the poles. Next, we resampled the data in the spectral dimension to a

[1] We found a GDAL bug when reading 8 bytes real values that is solved for version $\geq 2.3.0$ after our report to the developer. See https://trac.osgeo.org/gdal/wiki/Release/2.3.0-News and use this version or higher when manipulating MASCS data.

[2] PostgreSQL is a relational database management that controls the creation, integrity, maintenance and use of a database.

[3] PostGIS adds support for geographic objects in geographic information system and extends the database language with functions to create and manipulate geographic objects. PostGIS follows the Simple Features for SQL specification from the Open Geospatial Consortium (OGC).

common wavelength range[4] from 260 nm to 1052 nm with a 4 nm resolution (2 nm spectral sampling), resulting in 396 spectral channels. This approach slightly oversamples the original 4.77 nm spectral resolution and removes some points from the original 200-1050 nm range. The resulting data matrix is expressed in tabular form, with each row representing a single grid cell or pixel on the surface. The elements of each row are the spectral reflectance values from the VIS instrument at 396 (resampled) wavelengths. The dataset now has dimension $[N \times M]$ where N is the number of grid cells ($360 \times 180 = 64,800$) and M is the number of spectral features (396). Due to the incomplete coverage and data filtering, some grid cells are empty. After removing these empty cells, the size of the dataset is [55,399 × 396]. As an example, Fig. 7.1b shows the distribution of the normalized reflectance at a fixed wavelength (700 nm).

We used a recent dataset that had large-scale photometric corrections [9,10] and thus was almost free from observation geometry effects. However, extreme geometry is still present and is typically associated with high noise and some residual instrumental effects. Based on our empirical tests, we filtered out observations with an emission/incidence angle $\geq 80°$. We also calculated the median value per wavelength and per cell grid when constructing the global hyperspectral data cube and filtered out observations falling under the 2nd percentile and above 99.9^{th} percentile to clean some residual geometry effects. With this approach we create an effective noise filter while retaining enough observations to be able to analyze the entirety of the surface of the planet.

7.4 Learning from multivariate data

7.4.1 Dimensionality reduction: ICA

One way to discover salient patterns in a large set of multidimensional data is to perform dimensionality reduction. Two popular techniques for dimensionality reduction are Independent Component Analysis (ICA) and Principal Component Analysis (PCA).

PCA try to find a reduced-rank representation of the data and seeks for that best explains the variability of the data.

ICA is a case of blind source separation techniques and assumes that the observations are nonGaussian signals and statistically independent from

[4] MASCS VIS data have different wavelength sampling and part of the global Mercury campaign had different spectral binning.

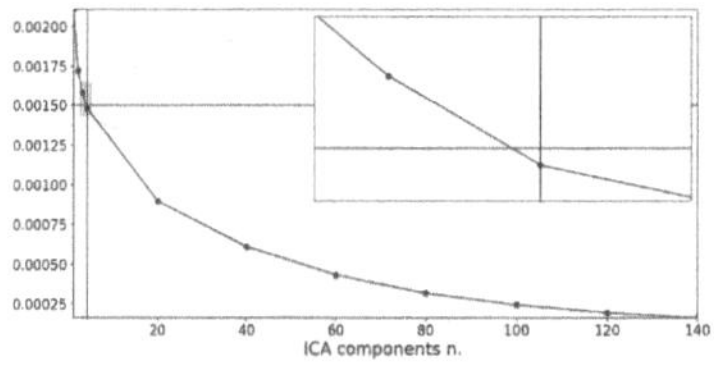

(a) ICA reconstruction error.

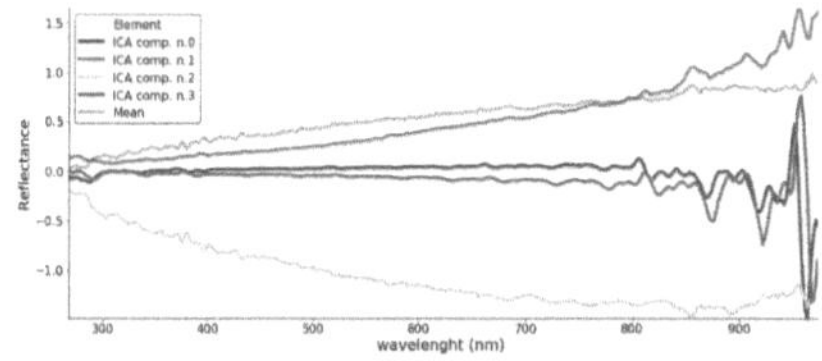

(b) ICA independent components.

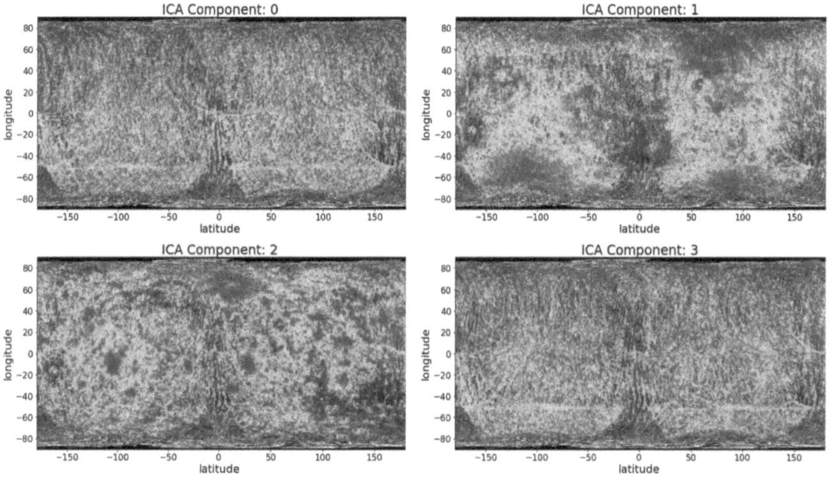

(c) ICA weights coefficients map.

Figure 7.2

each other [14,16]. ICA aims to find a set of new basis vectors as Principal Component Analysis, but with different assumptions and significance of the new basis. ICA search a basis where each vector is an independent component: if one think of a mix of superposed audio signals, the ICA basis will return a vector for each independent signal, like in the Blind Signal Separation techniques. ICA helps to find a representation of the data as independent subelements.

In practical terms, PCA helps to compress data and ICA helps to separate data.

The data cube obtained from previous section is then decomposed with ICA, trying to separate multivariate signal into additive subcomponents.

We represent the input data as $\hat{x} = (\hat{x}_1, ..., \hat{x}_n)^T$, where $\hat{x}_i$ is a vector of 396 spectral reflectance values for a given observation. ICA tries to find a linear transformation W so that $\hat{s} = W\hat{x}$, where $\hat{s} = (\hat{s}_1, ..., \hat{s}_k)^T$ is a vector of maximally independent components and $k \leq n$. W is also called the mixing matrix and has dimension $[f \times k]$ = [(data dimension)$\times$ (number of ICA independent sources)]. The original sources $\hat{s}$ can be recovered by multiplying the observed signals $\hat{x}$ with the inverse of the mixing matrix $W = A^{-1}$, also known as the unmixing matrix.

The goal is to find a number of independent components that can somehow be connected to geologic patterns observed on the surface. There is no "true number" of independent components based on the surface geology, so to choose k we use a reconstruction error threshold that represents the noise in the data. This value is chosen from the residual standard deviation when calculating the data median for each pixel in the surface $1° \times 1°$ grid (see 7.3) and represents the subpixel variation of the data.

The independent vectors s are function of the number k of component and by increasing it the reconstruction error decrease. Fig. 7.2a shows how the reconstruction error decreases with additional components and that the condition $||x - s(n)|| < 0.0015$ is true for $n \geq 4$. The final result is a compression of the data from [55,399 × 396] to [55,399 × 4].

In this application it is interesting to visualize the mixing weight matrix W, i.e., the linear operator that maps independent sources to the data $\hat{x}$. The weight coefficient for each component s represents how close each data point is to the corresponding independent component. Fig. 7.2c shows the weight coefficients maps, illustrating the spatial distribution of each component. Components 1 and 2 show well-defined clusters enriched (red (mid gray in print version)) and depleted (blue (dark gray in print version)) in the corresponding component. For components 0 and 3, the spatial distribution does not display any particular structure. This could be the result of some residual instrumental effects that introduce variance in the data.

7.4.2 Manifold learning

In the previous section, we reduced the entire dataset to a representation where each pixel in the surface grid was represented by 4 features computed using ICA instead of 396 spectral values (Fig. 7.2c). Before performing clustering on the dataset, we further reduced the dimensionality using manifold learning techniques. While it is technically possible to use manifold learning techniques to reduce the dimensionality without the ICA step, we applied ICA first because manifold learning algorithms do not computationally perform well for high-dimensional data. This is why they are usually used after a data compression step.

A number of supervised and unsupervised linear dimensionality reduction techniques exist, including but not limited to: Isomap, Locally

Linear Embedding (LLE), Hessian Eigenmapping, t-distributed Stochastic Neighbor Embedding (t-SNE), and Uniform Manifold Approximation and Projection (UMAP). See, e.g., the following review papers and the references therein: [11,22,31,34,35]. The linear algorithms most important aspect is that their results are easier to interpret than nonlinear algorithms, but by definition they could miss important nonlinear structures in the data.

Manifold Learning methods are a class of nonlinear dimensionality reduction techniques. They are based on the idea that the intrinsic dimensionality of many data sets is actually much lower than its actual dimensionality.

The linear dimensionality reduction techniques can be considered also dimensionality reduction techniques, because this terms embraces a broader set of algorithms. If we look at ICA and the Manifold Learning methods as a black box, they both compress input data. Their underlying assumption, the steps to find the lower dimension representation and the insight we gain from them are different, as we will see below.

The typical manifold learning problem is unsupervised: it learns the high-dimensional structure of the data from the data itself, without the use of predetermined classes [18].

UMAP and t-SNE are popular methods for manifold learning. At its core, t-SNE projects high-dimensional data in a space of lower dimensions. t-SNE initialization aims to model the probability distribution that represents similarities between neighbors (data points). It starts with a small random Gaussian distribution and for the first iterations the input probabilities are multiplied by an integer factor to 'exaggerate' the distribution. t-SNE converts similarities between data points to joint probabilities then tries to minimize the divergence between the joint probabilities of the low-dimensional embedding and the high-dimensional data. Original data points that are similar are modeled by nearby points in the low dimensional embedding. Dissimilar data points are modeled by distant points with high probability in the embedded space.

Compared to other techniques, t-SNE is particularly sensitive to local structure and can reveal the structure at many scales on a single map for data that lie in different manifolds or clusters. While linear methods such as Isomap and LLE unfold a single continuous low dimensional manifold, t-SNE will tend to extract clustered local groups of samples, since it focuses on the local structure of the data. For a complete overview of t-SNE we refer the reader to [31] and the references therein.

While powerful, t-SNE is extremely computationally expensive and can take several hours on million-sample data sets where PCA will take seconds to minutes. In addition, the downside of using a random initialization is repeatability: even if you use a fixed seed, there is no guarantee that the final results will be the same. This means that t-SNE can produce results with different initialization on the same data, so a good practice is to run t-SNE for multiple random seeds for the same dataset. Thus, we chose to apply UMAP [22] since this method offers a number of advantages over t-SNE, most notably increased speed with preservation of the global data structure.

We will give here an high level description of UMAP, trying not to make the description too complicated and not to fall into areas beyond this book. This can leave room for ambiguity and errors that are only our fault due to too much simplification of a complex topic. A rigorous mathematical background could be fund in [22] and nice visualization in the https://umap-learn.readthedocs.io/en/latest/how_umap_works.html.[5]

UMAP uses local manifold approximations, represent those with a local fuzzy simplicial set representations and uses their union to construct a topological representation of the high dimensional data. A simplicial complex is a set composed of points, line segments, triangles, and their n-dimensional counterparts. Fuzzy topology means that being in an open set is no longer a binary yes or no, but instead a fuzzy value between zero and one. The probability of being in a ball of a given radius for points will decay moving away from the center of the ball. Given some low dimensional representation of the data, a similar process can be used to construct an equivalent topological representation in the other direction (low to higher dimension). UMAP then searches for a low dimensional projection of the data that has the closest possible equivalent fuzzy topological structure to the high dimensional one.

UMAP follows the philosophy of t-SNE, but has a number of improvements such a different cost function and the absence of normalization of high- and low-dimensional probabilities. Further details can be found in the original paper, [22], and its reference python implementation, [21].

The two main hyperparameters controlling the representation learned by UMAP are *n_neighbors* and *min_dist*.[6] With *n_neighbors* one can adjust

[5] https://umap-learn.readthedocs.io/en/latest/how_umap_works.html.

[6] The interactive tutorial "Understanding UMAP" give some insight in how the hyperparameters influence UMAP. See [5] and https://pair-code.github.io/understanding-346umap.

the sensitivity to local or global structures in the data by constraining the size of the local neighborhood. Low values correspond to a focus on very local structures, possibly losing the big picture, while large values, corresponding to larger neighborhoods, may miss the fine structure. The best value to choose depends on the particular data structure and the desired range (local vs. global) the application is required to probe. Values can go from 0 (local) to the size of the data (global). *min_dist* regulates how points are packed together, providing the minimum distance apart that points are allowed to be in the low dimensional representation. Low values will result in more dense embeddings while larger values will result in more sparse embeddings.

Fig. 7.3a shows the density pair plots of the four ICA weight coefficients. This shows that the data manifold is a slightly asymmetric 4D cloud, with most of the data concentrated in the center. Fig. 7.3b shows the UMAP density plot on the ICA weight coefficients for various combinations of the two hyperparameters described above. The projected data topology follow the 4D manifold, with a central core of high density observations and a rarefied peripheral region. We chose to use $n\_neighbors = 4000$ and $min\_dist = 0.99$. In our particular application because local structures are not apparent at any level, so the only significant structure is the overall shape of the low dimension representation and how where the data points lies relative to each other.

After using UMAP to further reduce the representation of the dataset compressed using ICA, the final dimensionality of the dataset is $[55{,}399 \times 2]$.

7.4.3 Cluster analysis

After reducing the dimensionality of the data, our final goal is to identify surface regions with similar spectral properties that we can use to make geologic interpretations. As described in the Introduction, this is because geologic surfaces are generally classified into a few classes which are related with a geological taxonomy.

Cluster analysis comprises a series of techniques that attempt to group similar observations in groups or clusters. The idea is that observations belonging to a cluster are more similar to each other than to those in other clusters. Before clustering the data, we first standardized the data by subtracting the mean and scaling them to unit variance. This is a standard and often necessary step for any machine learning algorithm.

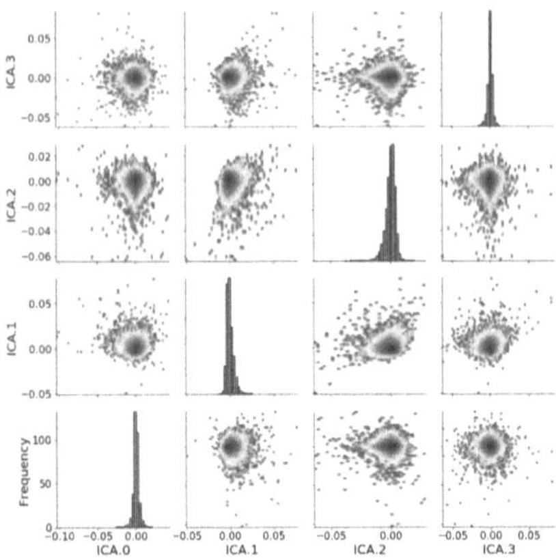

(a) ICA_4 weight coefficient density pair-plots.

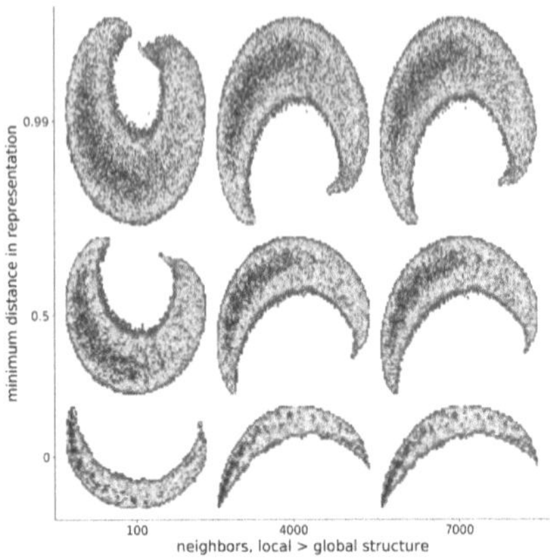

(b) UMAP on ICA_4 exploring hyperparameters space.

Figure 7.3

K-means clustering is a very popular clustering algorithm due to its computational efficiency and its many available implementations. K-means is a vector quantization technique, where an observation belongs to the cluster with the nearest mean (cluster center or cluster centroid), serving as a prototype of the cluster. K-means assumes that the variance of the distribution of each feature is spherical, that all features have similar variance, and that the prior probability for all k clusters is the same, i.e., each cluster has roughly equal numbers of observations.

Another family of clustering algorithms is known as hierarchical clustering algorithms, which build nested clusters by merging or splitting them successively. Hierarchical agglomerative clustering merges closer points together, irrespective of the final cluster class balance. The hierarchy of clusters is represented as a dendrogram as in Fig. 7.4d. The final tree has one unique cluster comprising all the samples at the root and clusters with only one sample at the leaves (bottom of the tree). The branch lengths represent the distance between the child clusters. Small gaps connect more similar clusters and big gaps connect more different clusters. The cluster distance is computed as the maximum Euclidean distance distances between all observations of the two clusters.

We choose the agglomerative clustering algorithm because it could be easily adapted to use different distance metrics and directly produce the tree like representation of the whole clusterization (the dendrogram). It is more

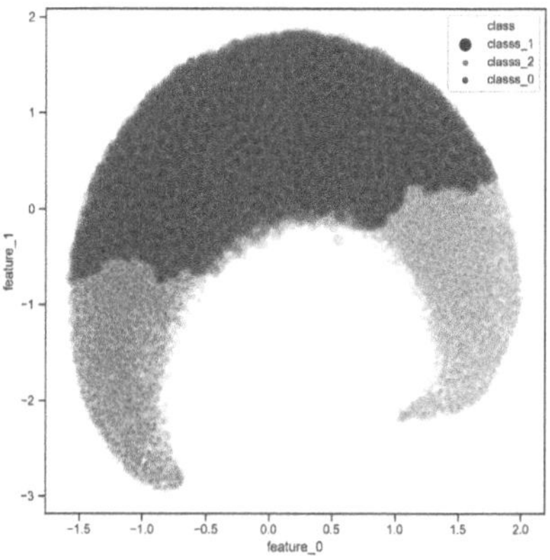

(a) Agglomerative Clustering labels on UMAP features.

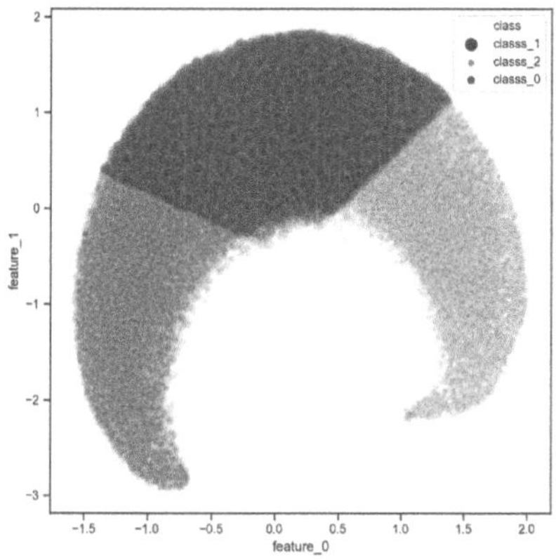

(b) K-Means labels on UMAP features.

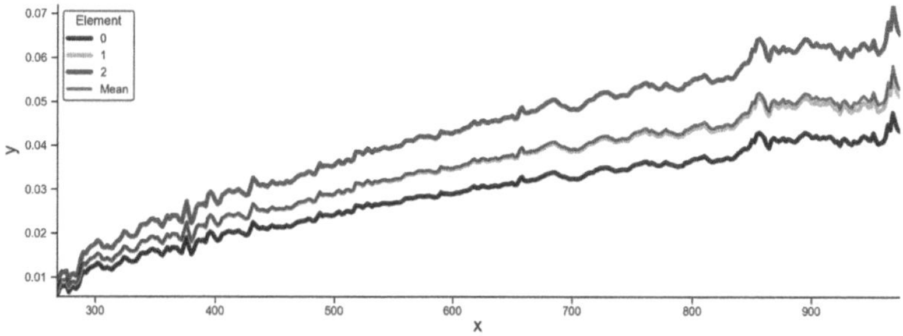

(c) Agglomerative Clustering centroids.

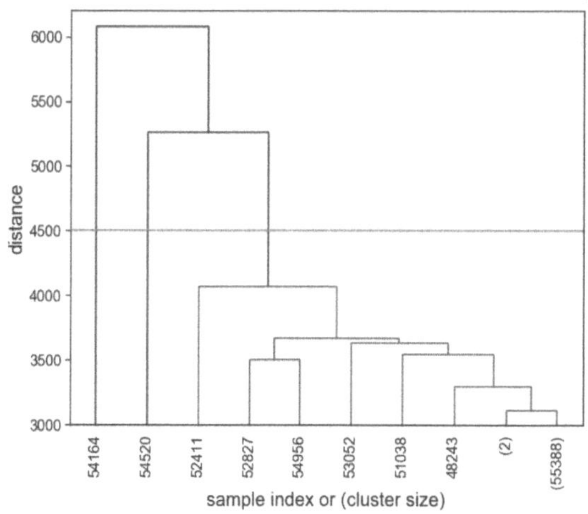

(d) Agglomerative Clustering dendrogram.

Figure 7.4

computationally intensive than the K-mean, but the latter gives directly a way to judge the correct number of cluster.

We used the reconstruction error threshold to choose the value of k, i.e., the number of clusters. Interested readers could explore other methods like the silhouette score[7] calculated using the mean intra-cluster distance and the mean nearest-cluster distance for each sample. Figs. 7.4a and 7.4b show the resulting cluster assignments in the UMAP embedded space when using agglomerative clustering and K-means clustering with $k = 3$ clusters.

The clustering methods we described are just a few of the available clustering techniques and methods may have better results for different datasets [32]. For example, when dealing with data containing clusters of similar density interleaved with low density zones (like clusters of data in some dimension with added noise), Density Based Spatial Clustering of Applications with Noise (DBSCAN) may give better results. DBSCAN separates core samples of high density regions and expands clusters from them.

7.4.4 Conclusion

The result of our analysis is a set of cluster maps such as in Fig. 7.5a. These maps can be directly compared with models (i.e. surface maximum temperature, roughness or age) and with expert-generated maps that are based on different methods of analysis described in the Introduction. Fig. 7.5b shows the finer clusters defined when splitting the data in 12 clusters. The structure is essentially the same as the 3-cluster case, indicating that this reduction does not appreciably affect the information content gained from the cluster map. The spatial distribution of the red (dark gray in print version) cluster in the north compares favorably with the location of younger volcanic smooth plains plotted in Fig. 7.5d. These plains are widespread on Mercury, but are more concentrated in the north and in the surroundings of the Caloris impact basin (30.5°N 189.8°W) and are most likely of volcanic origin. The smooth plains are younger than the Caloris basin, as indicated by the lower crater densities in those regions [7]. The basin floor is filled by geologically distinct plains implanted after the impact at the origin of the basin itself.

[7] See for example "Selecting the number of clusters with silhouette analysis on KMeans clustering" https://scikit-learn.org/stable/auto_examples/cluster/plot_kmeans_silhouette_analysis.html.

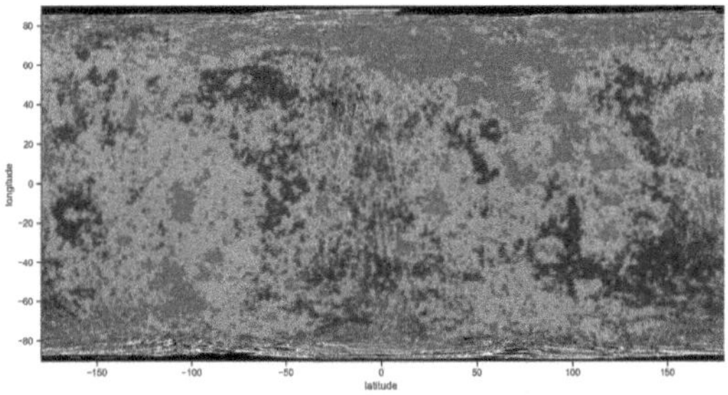

(a) Agglomerative Clustering 3 classes

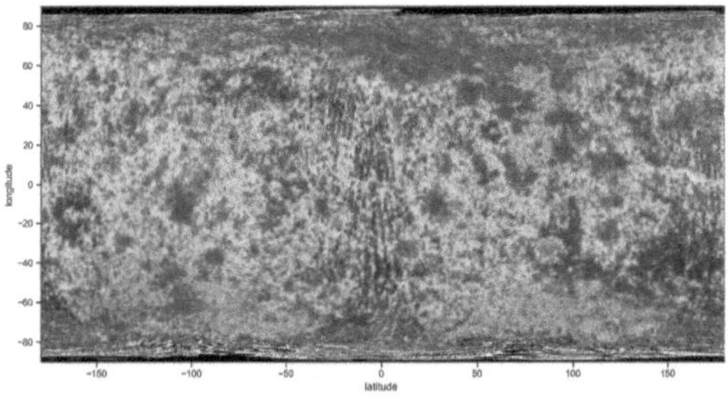

(b) Agglomerative Clustering 12 Classes

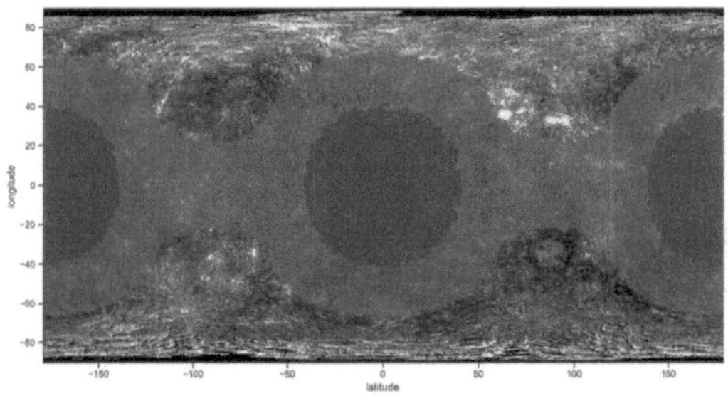

(c) Maximum surface temperature: red > 690 K, brown > 550 K, adapted from Vasavada (1999)

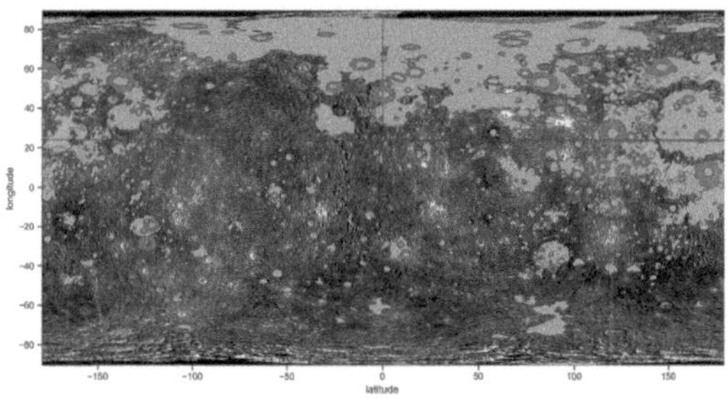

(d) Mercury smooth plains, adapted from Denevi et al. (2013)

Figure 7.5

The result of our works is the division of the surface in three classes or clusters, that capture some inherent properties of Mercury in the VNIR range as measured by MASCS instrument.

The red (dark gray in print version) clusters (Fig. 7.5a) are concentrated outside the hot poles, hinting at the possible role that thermal processing could play in modifying the spectra of the surface.

The representative spectra for those classes differ mostly for the spectral tilt, which is the ratio between longer (redder) and shorter (bluer) wavelengths, as shown in Fig. 7.4c. It is important to stress that this is a general property of Mercury MASCS spectra. They do not show any significant absorption bands, thus it is difficult to match the spectra with known mineral assemblages. This is even more pronounced taking the average of a great number of spectra, washing out single spectral characteristic. This is not always the case, i.e. Bandfield et al. [1] did found spectral classes on Mars and linked them to terrestrial basaltic and an andesitic composition, roughly related to the Martian dichotomy and older/younger surfaces.

In the case of Mercury, several factors are in play.

Mercury's 3:2 spin–orbit resonance with the Sun creates the characteristic surface temperature distribution shown in Fig. 7.5c, which maps the maximum temperature reached at each point of the surface, based solely on solar irradiation. There are two hot poles around 0° and 180°.

Thermal processing is known to tilt spectra to higher reddening, i.e., steeper spectra [19].

Solar wind irradiation also concurs to redden the spectra, even though the literature on this topic focused mostly on asteroids with high-Fe content, while Mercury's surface is extremely low in iron [28]. Mercury reflectance spectra are also extremely dark and amorphous carbon has been proposed as darkening agent [30].

To summarize, dimensionality reduction, manifold learning, and cluster analysis techniques are powerful tools to explore large sets of uncategorized planetary data. The workflow we presented could be easily extended to include additional instrument data, for example including data on the same geographical locations of chemical composition extracted from the X-Ray Spectrometer (XRS) [27,28]. While we focused only on reflectance data in this study, the technique of data fusion could further promote the discovery of scientifically-interesting patterns in the data, thus helping to understand complex physical mechanisms that could not be revealed by data coming from a single instrument.

Acknowledgment

The code and data used in this work is available as Python Jupyter Notebook on the github public repository MESSENGER-Mercury-Surface-Cassification-Unsupervised_DLR[8] funded under the European Union's Horizon 2020 research and innovation programme grant agreement No 871149. The Working Package "Machine Learning Solutions for Data Analysis and Exploitation in Planetary Science" aims to develop and implement an ambitious sustainable ML toolset, tailored for the needs of the planetary science community by analyzing a representative set of scientific cases.

References

[1] J. Bandfield, V. Hamilton, P. Christensen, A global view of Martian surface compositions from MGS-TES, Science 287 (March) (2000) 1626–1630, https://doi.org/10.1126/science.287.5458.1626.

[2] S. Besse, A. Doressoundiram, J. Benkhoff, Spectroscopic properties of explosive volcanism within the Caloris basin with MESSENGER observations, Journal of Geophysical Research: Planets 120 (12) (December 2015) 2102–2117, https://doi.org/10.1002/2015JE004819, ISSN 21699097.

[3] P. Bland, Crater counting, Astronomy & Geophysics (ISSN 1366-8781) 44 (4) (August 2003) 4.21, https://doi.org/10.1046/j.1468-4004.2003.44421.x.

[4] D.T. Blewett, N.L. Chabot, B.W. Denevi, C.M. Ernst, J.W. Head, N.R. Izenberg, S.L. Murchie, S.C. Solomon, L.R. Nittler, T.J. McCoy, Z. Xiao, D.M.H. Baker, C.I. Fassett, S.E. Braden, J. Oberst, F. Scholten, F. Preusker, D.M. Hurwitz, Hollows on Mercury: MESSENGER evidence for geologically recent volatile-related activity, Science (ISSN 0036-8075) 333 (6051) (Sep 2011) 1856–1859, https://doi.org/10.1126/science.1211681.

[5] A. Coenen, A. Pearce, Understanding UMAP, https://pair-code.github.io/understanding-umap/, 2019.

[6] B.W. Denevi, M.S. Robinson, D.T. Blewett, D.L. Domingue, J.W. Head III, T.J. McCoy, R.L. McNutt Jr., S.L. Murchie, S.C. Solomon, MESSENGER global color observations: implications for the composition and evolution of Mercury's crust, in: Lunar and Planetary Science Conference, 2009, pp. 1–2.

[7] B.W. Denevi, C.M. Ernst, H.M. Meyer, M.S. Robinson, S.L. Murchie, J.L. Whitten, J.W. Head, T.R. Watters, S.C. Solomon, L.R. Ostrach, C.R. Chapman, P.K. Byrne, C. Klimczak, P.N. Peplowski, The distribution and origin of smooth plains on Mercury, Journal of Geophysical Research: Planets 118 (5) (May 2013) 891–907, https://doi.org/10.1002/jgre.20075, ISSN 21699097.

[8] P. D'Incecco, J. Helbert, M. D'Amore, A. Maturilli, J.W.J. Head, R.L.R. Klima, N.R.N. Izenberg, W.E.W. McClintock, H. Hiesinger, S. Ferrari, Shallow crustal composition of Mercury as revealed by spectral properties and geological units of two impact craters, Planetary and Space Science 119 (2015) 250–263, https://doi.org/10.1016/j.pss.2015.10.007, ISSN 00320633.

[9] D.L. Domingue, M. D'Amore, S. Ferrari, J. Helbert, N.R. Izenberg, Analysis of the MESSENGER MASCS photometric targets part I: photometric standardization for examining spectral variability across Mercury's surface, Icarus (ISSN 0019-1035) 319 (February 2019) 247–263, https://doi.org/10.1016/j.icarus.2018.07.019.

[8] https://github.com/epn-ml/MESSENGER-Mercury-Surface-Cassification-Unsupervised_DLR.

[10] D.L. Domingue, M. D'Amore, S. Ferrari, J. Helbert, N.R. Izenberg, Analysis of the MESSENGER MASCS photometric targets part II: photometric variability between geomorphological units, Icarus (ISSN 0019-1035) 319 (February 2019) 140–246, https://doi.org/10.1016/j.icarus.2018.07.018.

[11] D.L. Donoho, C. Grimes, Hessian eigenmaps: locally linear embedding techniques for high-dimensional data, Proceedings of the National Academy of Sciences 100 (10) (May 2003) 5591–5596, https://doi.org/10.1073/pnas.1031596100, ISSN 0027-8424, 1091-6490.

[12] K.E. Vander Kaaden, F.M. McCubbin, L.R. Nittler, P.N. Peplowski, S.Z. Weider, E.A. Frank, T.J. McCoy, Geochemistry, mineralogy, and petrology of boninitic and komatiitic rocks on the mercurian surface: insights into the mercurian mantle, Icarus (ISSN 0019-1035) 285 (March 2017) 155–168, https://doi.org/10.1016/j.icarus.2016.11.041.

[13] V.E. Hamilton, H.Y. McSween, B. Hapke, Mineralogy of Martian atmospheric dust inferred from thermal infrared spectra of aerosols, Journal of Geophysical Research (ISSN 0148-0227) 110 (E12) (2005) 1–11, https://doi.org/10.1029/2005JE002501.

[14] T. Hastie, R. Tibshirani, J. Friedman, The Elements of Statistical Learning: Data Mining, Inference, and Prediction, second edition, Springer Series in Statistics, Springer-Verlag, New York, ISBN 978-0-387-84857-0, 2009.

[15] J. Helbert, A. Maturilli, M. D'Amore, M. D'Amore, Visible and near-infrared reflectance spectra of thermally processed synthetic sulfides as a potential analog for the hollow forming materials on Mercury, Earth and Planetary Science Letters 369–370 (May 2013) 233–238, https://doi.org/10.1016/j.epsl.2013.03.045, ISSN 0012821X.

[16] A. Hyvärinen, E. Oja, Independent component analysis: algorithms and applications, Neural Networks (ISSN 0893-6080) 13 (4) (June 2000) 411–430, https://doi.org/10.1016/S0893-6080(00)00026-5.

[17] R.A. Kerr, Who can read the Martian clock?, Science 312 (5777) (May 2006) 1132–1133, https://doi.org/10.1126/science.312.5777.1132, ISSN 0036-8075, 1095-9203.

[18] J.A. Lee, M. Verleysen, Nonlinear Dimensionality Reduction, Information Science and Statistics, Springer-Verlag, New York, ISBN 978-0-387-39350-6, 2007.

[19] A. Maturilli, J. Helbert, J. St. John, J. Head, W. Vaughan, M. D'Amore, M. Gottschalk, S. Ferrari, Komatiites as Mercury surface analogues: spectral measurements at PEL, Earth and Planetary Science Letters 398 (2014), https://doi.org/10.1016/j.epsl.2014.04.035, ISSN 0012821X.

[20] W.E. McClintock, M.R. Lankton, The Mercury atmospheric and surface composition spectrometer for the MESSENGER mission, Space Science Reviews 131 (1–4) (2007) 481–521, https://doi.org/10.1007/s11214-007-9264-5, ISSN 00386308.

[21] L. McInnes, UMAP: uniform manifold approximation and projection for dimension reduction — umap 0.5 documentation, https://umap-learn.readthedocs.io/en/latest/index.html, 2018.

[22] L. McInnes, J. Healy, J. Melville, UMAP: uniform manifold approximation and projection for dimension reduction, arXiv:1802.03426 [cs, stat], September 2020.

[23] P.-Y. Meslin, O. Gasnault, O. Forni, S. Schröder, A. Cousin, G. Berger, S.M. Clegg, J. Lasue, S. Maurice, V. Sautter, S.L. Mouélic, R.C. Wiens, C. Fabre, W. Goetz, D. Bish, N. Mangold, B. Ehlmann, N. Lanza, A.-M. Harri, R. Anderson, E. Rampe, T.H. McConnochie, P. Pinet, D. Blaney, R. Léveillé, D. Archer, B. Barraclough, S. Bender, D. Blake, J.G. Blank, N. Bridges, B.C. Clark, L. DeFlores, D. Delapp, G. Dromart, M.D. Dyar, M. Fisk, B. Gondet, J. Grotzinger, K. Herkenhoff, J. Johnson, J.-L. Lacour, Y. Langevin, L. Leshin, E. Lewin, M.B. Madsen, N. Melikechi, A. Mezzacappa, M.A. Mischna, J.E. Moores, H. Newsom, A. Ollila, R. Perez, N. Renno, J.-B. Sirven, R. Tokar, M. de la Torre, L. d'Uston, D. Vaniman, A. Yingst, M.S. Team,

Soil diversity and hydration as observed by ChemCam at Gale crater, Mars, Science 341 (6153) (September 2013), https://doi.org/10.1126/science.1238670, ISSN 0036-8075, 1095-9203.
[24] O. Namur, B. Charlier, Silicate mineralogy at the surface of Mercury, Nature Geoscience (ISSN 1752-0908) 10 (1) (January 2017) 9–13, https://doi.org/10.1038/ngeo2860.
[25] NASA, PDS: PDS3 standards reference, https://pds.nasa.gov/datastandards/pds3/standards/, 2009.
[26] C.D. Neish, D.T. Blewett, J.K. Harmon, E.I. Coman, J.T. Cahill, C.M. Ernst, A comparison of rayed craters on the moon and Mercury, Journal of Geophysical Research E: Planets 118 (10) (2013) 2247–2261, https://doi.org/10.1002/jgre.20166, ISSN 01480227.
[27] L.R. Nittler, R.D. Starr, S.Z. Weider, T.J. McCoy, W.V. Boynton, D.S. Ebel, C.M. Ernst, L.G. Evans, J.O. Goldsten, D.K. Hamara, D.J. Lawrence, R.L. McNutt, C.E. Schlemm, S.C. Solomon, A.L. Sprague, The major-element composition of Mercury's surface from MESSENGER X-ray spectrometry, Science (ISSN 0036-8075) 333 (6051) (September 2011) 1847–1850, https://doi.org/10.1126/science.1211567.
[28] L.R. Nittler, E.A. Frank, S.Z. Weider, E. Crapster-Pregont, A. Vorburger, R.D. Starr, S.C. Solomon, Global major-element maps of Mercury from four years of MESSENGER X-ray spectrometer observations, Icarus (ISSN 0019-1035) (February 2020) 113716, https://doi.org/10.1016/j.icarus.2020.113716.
[29] S. Padovan, N. Tosi, A.-C. Plesa, T. Ruedas, Impact-induced changes in source depth and volume of magmatism on Mercury and their observational signatures, Nature Communications (ISSN 2041-1723) 8 (1) (December 2017) 1945, https://doi.org/10.1038/s41467-017-01692-0.
[30] P.N. Peplowski, R.L. Klima, D.J. Lawrence, C.M. Ernst, B.W. Denevi, E.A. Frank, J.O. Goldsten, S.L. Murchie, L.R. Nittler, S.C. Solomon, Remote sensing evidence for an ancient carbon-bearing crust on Mercury, Nature Geoscience (ISSN 1752-0894) 9 (4) (March 2016) 273–276, https://doi.org/10.1038/ngeo2669.
[31] S.T. Roweis, L.K. Saul, Nonlinear dimensionality reduction by locally linear embedding, Science 290 (5500) (December 2000) 2323–2326, https://doi.org/10.1126/science.290.5500.2323, ISSN 0036-8075, 1095-9203.
[32] Rui Xu, D. Wunsch, Survey of clustering algorithms, IEEE Transactions on Neural Networks (ISSN 1941-0093) 16 (3) (May 2005) 645–678, https://doi.org/10.1109/TNN.2005.845141.
[33] A. Sprague, K. Donaldson Hanna, R. Kozlowski, J. Helbert, A. Maturilli, J. Warell, J. Hora, Spectral emissivity measurements of Mercury's surface indicate Mg- and Ca-rich mineralogy, K-spar, Na-rich plagioclase, rutile, with possible perovskite, and garnet, Planetary and Space Science 57 (3) (March 2009) 364–383, https://doi.org/10.1016/j.pss.2009.01.006, ISSN 00320633.
[34] J.B. Tenenbaum, V. de Silva, J.C. Langford, A global geometric framework for nonlinear dimensionality reduction, Science 290 (5500) (December 2000) 2319–2323, https://doi.org/10.1126/science.290.5500.2319, ISSN 0036-8075, 1095-9203.
[35] L. van der Maaten, G. Hinton, Visualizing data using t-SNE, Journal of Machine Learning Research (ISSN 1533-7928) 9 (86) (2008) 2579–2605.
[36] A. Vasavada, Near-surface temperatures on Mercury and the Moon and the stability of polar ice deposits, Icarus 141 (2) (October 1999) 179–193, https://doi.org/10.1006/icar.1999.6175, ISSN 00191035.
[37] F. Vilas, D.L. Domingue, J. Helbert, M. D'Amore, A. Maturilli, R.L. Klima, K.R. Stockstill-Cahill, S.L. Murchie, N.R. Izenberg, D.T. Blewett, W.M. Vaughan, J.W. Head, Mineralogical indicators of Mercury's hollows composition in MESSENGER color observations, Geophysical Research Letters 43 (4) (February 2016) 1450–1456, https://doi.org/10.1002/2015GL067515, ISSN 00948276.

CHAPTER 8

Mapping storms on Saturn

Ingo P. Waldmann
University College London, London, United Kingdom

8.1 Introduction

Planetary space missions, such as Juno, Cassini-Huygens, Mars Reconnaissance Orbiter and Venus Express produce high-quality data sets that transform our understanding of solar system bodies. Yet, the sheer amount of data obtained over the lifetime of a mission prohibits detailed 'by hand' analyses of each spectrum and each spatial pixel. This is particularly true for the characterization of planetary atmospheres.

To date, atmospheric analysis of gas-giant planets in our solar system is often done by either one of two methods: 1) dividing the spectral flux (I/F) of particular spectral channels from another to identify the existence of spectral features; 2) performing full radiative transfer (RT) calculations of planetary spectra at given latitude/longitude locations. Though complementary, both methods have significant drawbacks. By design, the I/F method can only provide crude estimates of the existence of atmospheric features, is limited by the choice of spectral bands used and neglects important information in other spectral regions. On the other hand, a full RT calculation will provide detailed quantitative results on the atmospheric make-up but is very computationally intensive. This limits a classical RT analysis to typically a handfull of longitude/latitude points on a planet. Hence, neither technique is well suited to accurately identify and map large-scale atmospheric features, in particular if these features span multiple data sets.

The large body of planetary science data lends itself to the recent advances in data analysis and pattern recognition. In this example, we focus in particular on hyperspectral imaging instruments such as the Visible and Infrared Mapping Spectrometer (VIMS) on the Cassini-Huygens, the Compact Reconnaissance Imaging Spectrometer (CRISM) on the Mars Reconnaissance Orbiter or the VIRTIS instrument on Venus Express, to name a few. The use of machine learning and, in particular, deep learning is a very active field of research and has been applied successfully in various areas of planetary science (e.g. [1–6]). Similarly, hyperspectral image

Machine Learning for Planetary Science
https://doi.org/10.1016/B978-0-12-818721-0.00017-3
Copyright © 2022 Elsevier Inc.
All rights reserved.

analysis of aerial images for commercial land use classification has become standard practice in recent years (e.g. [7,8]). For a recent review we refer the reader to [9].

8.1.1 Cassini-Huygens and ammonia clouds

For the purpose of this example, we follow Waldmann and Griffith [10] and focus on hyperspectral images obtained by the VIMS instrument on board the Cassini-Huygens mission. The Cassini-Huygens mission was a NASA/ESA spacecraft the Saturnian system for 13 years until September 2017. The VIMS instrument [11] is a 2-channel mapping spectrometer, which obtains spatially resolved spectra across a 64×64 pixel array.

One of the prevalent mysteries of Saturn is the near total absence of ammonia clouds. Clouds and aerosols provide a unique insight into the chemical and physical processes of gas giant planets. Mapping and characterizing the spectral features indicative of the cloud structure and composition enables an understanding of a planet's energy budget, chemistry and atmospheric dynamics (e.g. [12–15]). Baines et al. [16] (B09 hereafter) reported a rare detection of a fresh ammonia condensate cloud in the vicinity of a dark storm in the southern hemisphere of Saturn. This feature, in the shape of an inverted S-shape, was largely found through visual inspection of the data, see Fig. 8.1. Though successful in this instance, such an approach limits us to the most prominent and strongest features present. The question remained whether this feature was an isolated case or whether undetected, similar features exist in data elsewhere.

Here we revisit the data presented by B09 and use the VIMS near-infrared channel in our analysis, which spans 256 contiguously sampled band-passes ranging from 0.85 to 5.1 μm with constant $\delta\lambda = 0.016$ μm. Hence each data cube (or hyperspectral image) had dimensions of $64 \times 64 \times 256$, amounting to 4096 individual spectra.

The data were obtained on February 9th 2008 and are available through the Planetary Data System[1] with data cube V1581233933 containing the "S" shaped ammonia ice feature, hereafter referred to as the training cube (TC).

8.2 Exploratory principal component analysis

Before embarking on more complex data analysis, it is often useful to conduct an initial exploration of the feature-space present in the hyperspectral

[1] https://pds.nasa.gov.

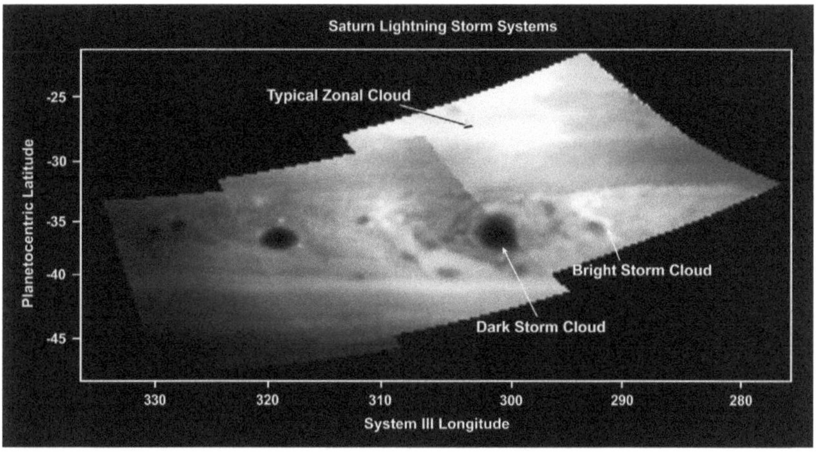

Figure 8.1 Original publication of the inverted "S" shape bright storm cloud feature in the vicinity of a dark storm in Saturn's southern hemisphere. Figure from [16].

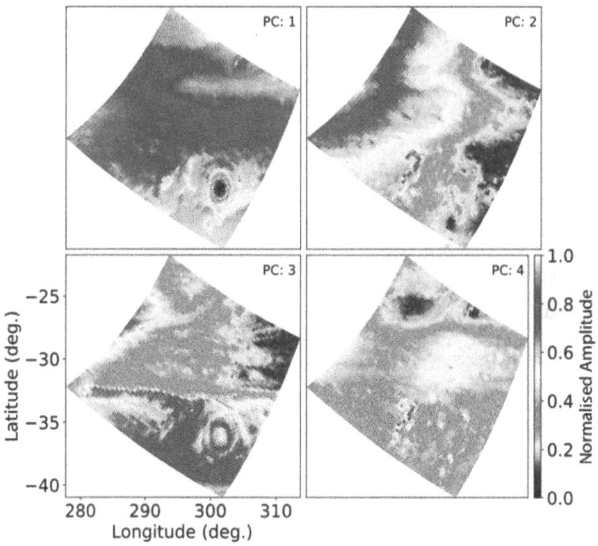

Figure 8.2 Spatial map of the principal component analysis of the 0.88 - 1.66 μm spectral region. Shown are the first four strongest principal components (PC). PC 1 is dominated by the dark storm, providing most variance to the spectral map. The 2nd to 4th PCs show the ammonia ice S-shape feature. Each principal component map is normalized to unity for clarity.

data. The end goal is to identify and accurately map distinct classes of spectral clusters in hyperspectral image data of Saturn. Should distinct spectral

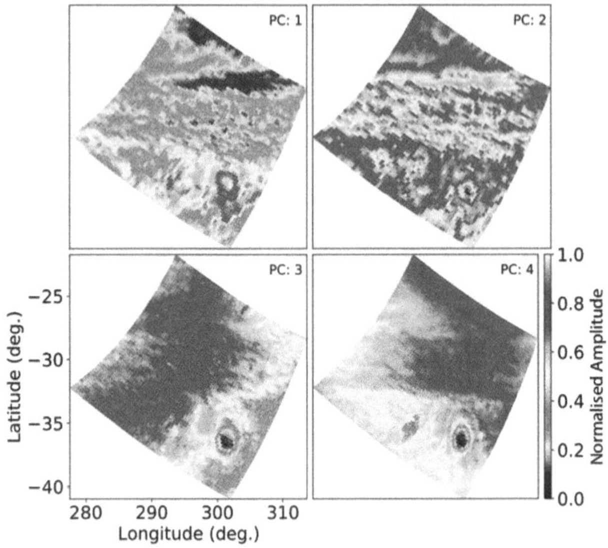

Figure 8.3 Spatial map of the principal component analysis of the full wavelength range of the instrument. Clearly visible are the dark storm, the extended SR feature and the high-altitude CH_4 clouds.

classes exist in the data, they will contribute to the variance of the data cube and should be visible as individual principal or independent components. Algorithms such as Principal Component Analysis (PCA, Jolliffe [17]) find a linear transformation from the correlated observed signals to a set of uncorrelated signals. Such a linear transformation is always possible and easily achieved using, for example, singular value decompositions (SVDs). Hence, should uncorrelated and separable features exist in the data, they may be separable to some degree by PCA into individual components [18]. In a recent analysis, Griffith et al. [19] has shown that PCA analysis of hyperspectral images of Titan can unveil structural information that otherwise remains hidden in spectral noise.

Whilst PCAs based on SVDs are the easiest to implement and sufficiently fast for the data set at hand, this is not always the case. As most features of interest will be contained in the first few principal components only, we often have no need to compute a full decomposition. Methods such as kernel-PCA (e.g. [20]) provide more efficient means of calculating the top components only.

By performing a simple principal component analysis, we can immediately see that clearly separated structures exist in the data. In Figs. 8.2 & 8.3

we show the first four principal components of the 0.88 - 1.66 μm spectral region and the full wavelength range respectively. Both maps clearly show the presence of significant structure surrounding the dark circular storm feature towards the bottom right corner of the image. This extended structure we attribute to a turbulent storm region (SR) surrounding the dark storm and is most notable in the second component in Fig. 8.3. The dark storm itself is most prominent in the 1st component and hence contributes the highest amount of variance to the image. The "S" shaped NH_3 ice feature (B09) on the other hand is clearly visible in the second to fourth components of Fig. 8.2. Spectroscopically, the shorter wavelength regions probe the albedo of the clouds and are hence most sensitive to the high altitude atmospheric variations.

In many instances, a PCA analysis may yield the desirable outcome and reveal the data's feature space. In the case of a single data set, this may indeed be sufficient as exploratory analysis. However, in the case of multiple data sets, it may be difficult to directly compare PCA results as the underlying features and noise properties may change from data set to data set. Here, it may be advantageous to train a neural network to identify the desired feature set, in our case the NH_3 ice feature, and to then use it to search for similar signatures in other data cubes. In the following section we will describe such an approach.

8.3 A deep learning approach

As the preliminary PCA analysis confirmed, our data is rich in distinct features and we can now build neural network able to identify them in the data.

The neural network approach proposed by Waldmann and Griffith [10][2] takes into account both the correlations between adjacent spectra as well as spatial correlations of the mean pixel map. By analyzing both spectral and spatial information, we can take into account the morphological and spectral signatures of atmospheric features on Saturn. In other words, a dark storm, for example, will have a distinct spectrum and spatial morphology that correlate together. By including these spatial-spectral correlations, the neural network will take all possible information available into account. Finally, the output of both spectra and spatial branches is fed into a fully

[2] Data and code are freely available at https://github.com/ucl-exoplanets.

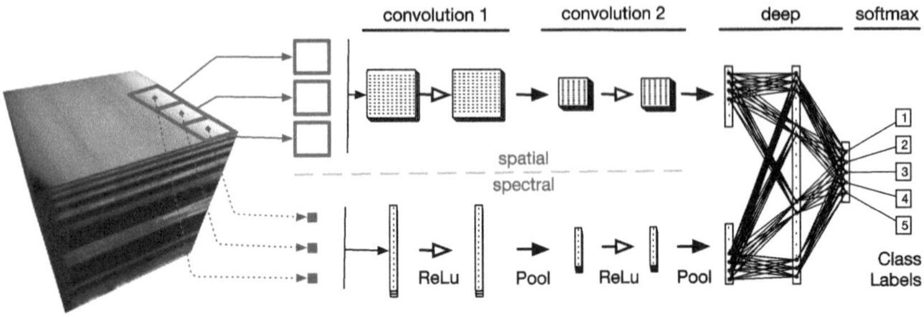

Figure 8.4 Left: Cassini/VIMS data cube. Blue (mid gray in print version) dots indicate the central pixel at which the spectrum is extracted. The red (mid gray in print version) squares indicate the spatial patches centered on the central spectral pixels. Middle: Two convolutional layers for the spectral and spatial information. Note that these are analyzed separately in parallel. Right: Fully connected deep neural network combining spatial and spectral convolutional outputs and mapping them to the cluster labels. The white and black arrows indicate ReLu and Pool layers respectively for both branches.

connected layer which links to the cluster labels via a logistic regression layer. Fig. 8.4 shows a schematic of the full network.

This architecture is able to identify faint features in hyperspectral images without underlying physics/model assumptions and, once trained on a given feature, able to search for it across unseen data sets. In more detail, the algorithm consists of two parts:

1. a preprocessing and spectral clustering stage to identify an initial feature set;
2. a double-stream deep convolutional neural network learning to map the feature set to unseen data.

8.3.1 Preprocessing and prelabeling

Each Cassini/VIMS data cube consists of two spatial and one spectral axis. The spatial dimensions of the cube are 64×64 pixels and 256 wavelength points. As input to our neural network, we require spectra/spatial pairs. These are generated as follows: For every pixel, the full spectrum is extracted, i.e. $S = 256$, where S denotes the spectral dimension of the pixel. Surrounding the central pixel, we also extract the spatial information. The patch size is defined by the height (H) and width (W), which we later determined to be 21 pixels, i.e. 20 pixels surrounding the central pixel. Here the spatial patch is the mean over all wavelengths. We chose the mean image in this case as it best represented variations due to the storm region and the quieter surrounding regions. We extract spatial/spectral pairs for all

4096 pixels. Note that spatial patches are overlapping from adjacent pixels, meaning that spatial information enters the calculation more than once but spectral information does not. This ensures that correlations between spectral and spatial information can be learned adequately.

In order to train the neural network, we must obtain labeled data. This can be obtained in two ways: 1) labeling individual pixels by hand to mark predominant features or 2) use a statistical clustering algorithm to obtain estimated labels on which the network can learn.

Here we use spectral clustering on the data cube initially presented by Baines et al. [16] to obtain the NH_3 signature. Spectral clustering [21] is a versatile clustering algorithm suited for high-dimensional vector spaces. Unlike with more traditional clustering algorithms such as k-nearest neighbors or k-means, spectral clustering is not dependent on the convexity of the individual cluster sets (i.e. the cluster distribution does not need to follow a convex shape) and can identify highly nonconvex clusters. Similarly, DBSCAN [22] can be used in these cases and yields comparable results. For spectral clustering, we follow Yu and Shi [23], Ng et al. [24] and use the sklearn [25] implementation. We define the symmetric graph Laplacian,

$$\boldsymbol{L}_{sym} = \boldsymbol{D}^{-1/2}\boldsymbol{L}\boldsymbol{D}^{-1/2} \tag{8.1}$$

where $\mathcal{L} = \boldsymbol{D} - \boldsymbol{A}$, $\boldsymbol{A}$ is the affinity matrix and $\boldsymbol{D}$ is the degree matrix. The elements of the affinity matrix are computed using

$$a_{i,j} = \exp\left(-\frac{1}{2\sigma^2} d^2[\boldsymbol{x}_{\lambda,i}, \boldsymbol{x}_{\lambda,j}]\right) \tag{8.2}$$

where i, j are the spatial indices of the data and d^2 is the L^2 norm or Euclidian distance given by

$$d_{i,j} = \sqrt{\sum (\boldsymbol{x}_i - \boldsymbol{x}_j)^2} \tag{8.3}$$

and σ the standard deviation of the distances, d. The degree matrix is given by

$$\boldsymbol{D} = \mathrm{diag}\left(\sum_j a_{i,j}\right). \tag{8.4}$$

Once the graph Laplacian is calculated, we decompose $\mathcal{L}_{sym}$ into its eigenvalues and eigenvectors and rank them from smallest to largest eigenvalue.

We estimated the likely number of clusters present in the data by using the eigenvalue heuristic [26]

$$N_{cluster} = \max(\epsilon_{i+1} - \epsilon_i) \qquad \text{for } i = [1, 2, ..., N_\epsilon - 1], \tag{8.5}$$

which yielded five distinct clusters. We now use k-means clustering on the matrix of eigenvectors to divide the data into $N_{cluster}$ labeled partitions. This provides us with training labels for each spatial coordinate (i, j). The resulting clusters are relatively well balanced apart from the inner storm region due to its comparatively small spatial extent. Since the area of interest is the potential ammonia upwelling surrounding this region but not the inner storm region itself which is geographically small. We do not mitigate for the class imbalance pertaining to the inner storm region. Since these cluster labels are estimated, some uncertainty in the cluster edges may persist. To mitigate cluster-edge uncertainties, we prune the edges and discard any pixel that lie along a cluster boundary. Similarly, we discard frame-edge pixels to avoid potential additional noise at the detector edges. In other words, the central pixel is always ≥21 pixels away from the detector edge. Finally, the remaining data-label pairs are randomly split into 70% training and 30% validation sets, but ensuring that there is no spatial overlap between training and validation sets for the central pixel. Given the relatively small data set, we do allow overlap for the pixels surrounding the central pixel. In larger data sets, any overlap can be avoided.

8.3.2 Neural network

The NN contains two branches, a spatial and a spectral one. The spectral branch takes each labeled spectrum and trains a two layer CNN using ReLu activation functions and two pooling layers. As previously mentioned, surrounding each spectrum, we compute a 21×21 spatial image by averaging the spectral cube along the spectral axis. This is the input for the spatial branch which otherwise follows the same NN architecture as the spectral branch, see Fig. 8.4. Both CNN branches are then connected to a fully-connected layer and the trained cluster labels.

Convolutional Neural Networks, also known as translation invariant networks, have been designed specifically for image analysis. Each CNN layer contains three separate stages: 1) Convolution, 2) nonlinear transform (ReLU) and 3) downsampling (pooling). In the convolution stage, the 2D image is convolved with a set of 'filter functions' or kernels. These filter functions are continuously learned from the data during training and provide the decomposition into spatial features (e.g. edges, blobs, etc.). We now convolve each filter (computing the dot product) with the input image to obtain the activation map for each filter. In the second stage, we apply a nonlinearity transformation to the activation maps. It is now standard practice to use rectifier linear units (ReLUs) or leaky ReLU as the CNN nonlinearity. We here use the classical ReLU which is given as $f(x) = \max(0, x)$ and effectively removes all negative entries of the activation map, increasing the nonlinearity of the transform. Finally, we downsample in size, or 'pool', the activation maps. Here we use simple 2×2 max-pooling, where the maximum value of a 2×2 grid is retained. This procedure is now repeated for the second CNN layer.

The CNN presented here is classical in all aspects apart from the inclusion of two separate and simultaneous streams that treat the spatial and spectral data. Due to the intrinsic correlations in the spectral branch, we opt for a 1D CNN architecture instead of fully connected layers only. We refer the interested reader to the standard literature on neural networks and CNNs (e.g. [18,27,28]). The neural network has been implemented in Tensorflow [29]. The input data on which training and subsequent classification is performed is denoted by

$$\boldsymbol{x} \in \mathbb{R}^{H \times W \times S} \tag{8.6}$$

where $\boldsymbol{x}$ is a tensor of dimensions H, W and S, denoting the height, width and spectral dimension of the data cube. Here we define tensors and matrices by lower and upper-case bold letters respectively. For ease of notation, we define the spectral-only input data as $\boldsymbol{x}_\lambda = \boldsymbol{x}^s$ and spatial-only as $\boldsymbol{x}_\phi = \boldsymbol{x}^{(H \times W)}$. As shown in Fig. 8.4, the CNN features two convolution and pooling layers ($l \in 1, 2, \ldots, L$) for each spatial and spectral branch. These can be defined as follows

$$\boldsymbol{y}_i^l = P_{\phi,\lambda} \left[\sum_j^{N^{l-1}} \beta(\boldsymbol{y}_j^{l-1} \otimes \boldsymbol{w}_{i,j}^l + \boldsymbol{b}_i^l) \right] \tag{8.7}$$

where $\boldsymbol{y}_i^l$ is the CNN output at layer l and the feature map i and $\mathrm{P}_{\phi,\lambda}$ the pooling operator. For the spatial and spectral branches, $\boldsymbol{y}_i^l$ constitutes a 2D and 3D tensor ($\boldsymbol{y}_{\lambda,i}^l \in \mathbb{R}^{S^l \times B^l}$ and $\boldsymbol{y}_{\phi,i}^l \in \mathbb{R}^{H^l \times W^l \times B^l}$) respectively with B denoting the individual training batch size. $\boldsymbol{y}_j^{l-1}$ is the feature map of the previous layer and $\boldsymbol{w}_{i,j}^l$ the corresponding kernel of size K^l, and $\otimes$ designates linear convolution. We define the biases as $\boldsymbol{b}_i^l$ and note that $\boldsymbol{y}^0 = \boldsymbol{x}$. The activation function, β, is defined as a linear rectifier (ReLu) unit and a max-pooling has been used for both layers and branches.

After both convolution layers, the spatial and spectral branches are combined to form the input of a fully connected network. We define the fully connected part of the neural network as follows

$$z = \beta\{W^{c} \cdot \beta[W^{c-1} \cdot (y_{\phi}^{L} \odot y_{\lambda}^{L}) + b^{c-1}] + b^{c}\} \tag{8.8}$$

where $\odot$ denotes the concatenation of y_{ϕ}^{L} and y_{λ}^{L}, the outputs of the final convolution layers. The fully connected layer indices are given by $c \in 1, 2, \ldots, C$ and W and b are the weight matrices and biases respectively. The number of neurons per layer is defined by the hyperparameter M^{c}. For simplicity, we refer to the set of free parameters as $\vartheta = \{w^{l}, b^{l}, W^{c}, b^{c}\}$.

Finally, we map z to our training labels, θ, using a softmax regression layer. To train, we now minimize the cross-entropy of the system.

$$Cost = -\frac{1}{n}\sum_{x}\left[\theta \ln(a) + (1-\theta)\ln(1-a)\right] \tag{8.9}$$

where n is the total number of training data (or batch size), θ the vector of training labels and $a = \varsigma(z)$, where $\varsigma(.)$ is a sigmoid activation function.

8.3.3 Training and hyperparameter optimization

The neural network is trained for 20,000 iterations with mini-batch sizes of 100 spatial/spectral pairs at a time. We train on the training set only and use the validation set as unseen data to check the network's ability to generalize over new data and to gauge any overfitting. Overfitting would generally be observed in systems with decreasing cross-entropy but static or decreasing validation accuracy. To mitigate overfitting, we employ a 30% dropout rate across all free variables [33], ϑ, and a relatively slow learning rate of 0.001. We find that convergence in cross-entropy is usually observed around 15,000 training steps or less with a monotonically decreasing cross-entropy and increasing test and validation set accuracy, indicating good convergence and no overfitting. Training takes 60 minutes on a 6-core Intel Xeon E5 (3.5 Ghz) CPU or ~5 minutes on a Nvidia Tesla K40 GPU.

In order to optimize the size of the neural network, we ran a grid of 1000 hyperparameter sets. We define the hyperparameters as: the size of the spatial-patch surrounding the central pixel (red (mid gray in print version) square in Fig. 8.4, H and W), the number of feature maps for both spatial

Table 8.1 Summary of neural network hyperparameters and their optimization ranges.

Hyperparameters	Value	Range
H, W	$21, 21$	3 - 50
N^1, N^2	$15, 40$	3 - 50
K^l_λ, K^l_ϕ	4×1, 4×4	1 - 10
P_ϕ, P_λ	2×2, 2×1	1 - 10
M^1, M^2	1024, 10	10 - 2048

and spectral branches (N^l_ϕ and N^l_λ), the kernel size (K^l_ϕ and K^l_λ), the downsampling pool size (P_ϕ and P_λ) and the layer size of the fully-connected network (M^c). We used the validation loss as metric during the hyperparameter optimization. We define a symmetric square of 21×21 pixels (H, W) as input, i.e. effectively 20×20 pixels surrounding the central pixel. We further define $N^1 = 15$ and $N^2 = 40$ feature maps for both spatial and spectral branches for the first and second layers respectively. Both layers have kernel sizes of $K^l_\phi = 4 \times 4$ and $K^l_\lambda = 4 \times 1$ for the spatial and spectral branches respectively. We use a factor 2 downsampling with pooling sizes $P_\phi = 2 \times 2$ and $P_\lambda = 2 \times 1$ and a stride length of 2. We summarize the hyperparameter values and their ranges during hyperparameter optimization in Table 8.1.

8.3.4 Classification validation

Once fully trained, it is important to investigate the training and prediction accuracies on the training cube (TC). Before training, the TC was subdivided into 70% training data and 30% test data. The test data is not shown to the algorithm during training and is used to independently verify the classification accuracy and provide a diagnostic against overtraining. In the case of overtraining, the classification accuracy of the training set will steadily increase over the training period, whereas the test set accuracy remains unchanged or decreases. In this case, the neural network is memorizing the training set rather than learning how to generalize. Fig. 8.5 shows the training (red (dark gray in print version)) and test (purple (light gray in print version)) accuracies as function of training duration. Both curves are monotonically increasing and show no signs of overfitting. The training and validation sets reach $\sim 95\%$ and $\sim 90\%$ classification accuracies. Furthermore, Fig. 8.5 shows the cross-entropy (loss function) of the neural network during training.

After the neural network has been trained, we further verify the classification accuracy by testing the NN on a previously unseen test set, consisting

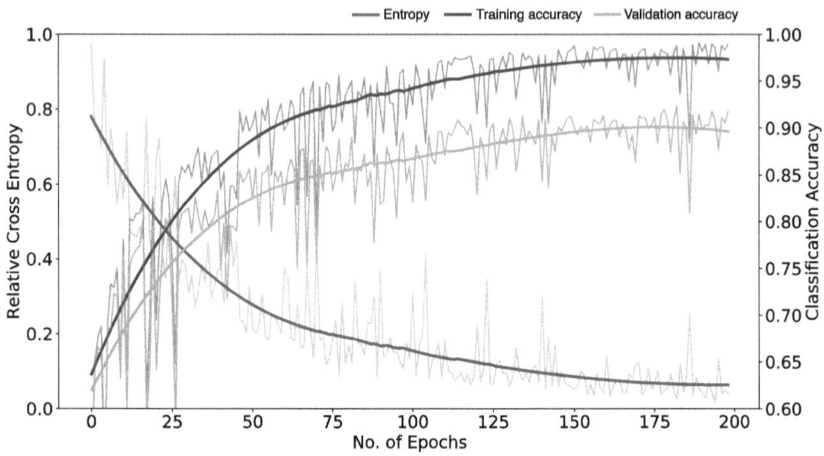

Figure 8.5 Left Axis/Blue (mid gray in print version): Relative cross entropy as function of training duration. Right axis: Training (red (dark gray in print version)) and Validation (purple (light gray in print version)) set accuracy as function of training duration. Bold lines present a smoothed average.

of a re-sampled, "synthetic" data set. We refer to this data set as re-sampled training cube (RTC). The RTC was generated by re-sampling the training cube in such a way as to preserve the statistical properties of the data but to appear as a new, unseen data to the NN. In the absence of definitive ground-truth labels in this study, we test the robustness of the NN to identify the original set of labels after re-sampling the data. It hence tests the relative robustness of the NN rather than the ground-truth classification accuracy. In cases where labeled data exists, through hand-labeling or otherwise, we suggest that these labels are used as test data.

First, the TC was divided according to the labels obtained by the spectral clustering. Following the previous notation, we refer to the TC data pertaining to a label as $\boldsymbol{x}_\theta$, where θ is the index of the label. Within each label subset, we randomized the order of each element of $\boldsymbol{x}_\theta$ to break any close spatial connections in the data. Now we re-sample each training set, both in the spatial $\boldsymbol{x}_{\theta,\phi}$ and spectral $\boldsymbol{x}_{\theta,\lambda}$ dimensions. For the spatial dimension, we transpose $\boldsymbol{x}_{\theta,\phi}$, which is equivalent to a 90 degree rotation. This rotation preserves the underlying properties of the data but appears as new, unseen data to the NN. The spectral information of each element in $\boldsymbol{x}_{\theta,\lambda}$ was replaced with the average of the two adjacent spectra in the RTC. Given that all spectral positions in the RTC are randomized, the newly generated spectra will effectively sample from the distribution of the cluster.

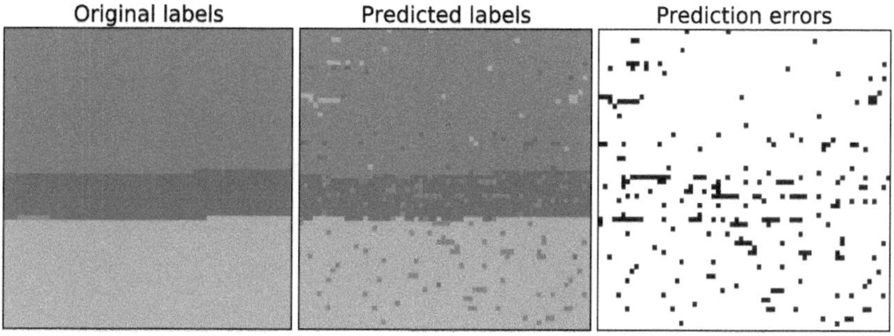

Figure 8.6 Left: Re-sampled training cube (RTC). Middle: Neural Network predicted labels. 94% of all pixels were correctly identified. Right: Error-matrix showing the wrongly identified pixels.

The NN is now run to detect cluster labels on RTC using spectral and spatial information, and additionally with spectral or spatial information only. Fig. 8.6 shows the RTC classification for spatial and spectral data. The left shows the 'ground-truth' labels, middle the predicted labels by the NN and right is the error matrix of pixels missclassified. The NN achieves a 93% classification accuracy on the test data, which is in agreement with the training/validation data classification accuracies. We now run the classification for spatial-only and spectra-only cases by setting $\boldsymbol{x}_{\theta,\lambda} = 0$ and $\boldsymbol{x}_{\theta,\phi} = 0$ respectively. This results in very poor cluster label recognition by the neural network (48% and 32% respectively). However we note that this is to be expected for models that are trained on both, spatial and spectral pairs. Suppressing half of the network's information results in erroneous predictions. Hence, if the user wants to use, say, spectral data only for their classification, the NN should be re-trained on spectral-only data.

8.4 Saturn's feature map

We apply our algorithm to the training cube, in which we identify five clearly distinguishable clusters of spatial/spectral features. These are broadly consistent with the PCA variance analysis.

The left of Fig. 8.7 presents the spatial extent of the TC and a map of the spectral features, while the right hand side shows the spectral characteristics of the five spectral clusters identified. Here the blue (mid gray in print version) region corresponds to a large stormy region (SR) surrounding the central dark storm (purple/green (dark gray/mid gray in print version)) and label 1 denotes the center of the "S" feature.

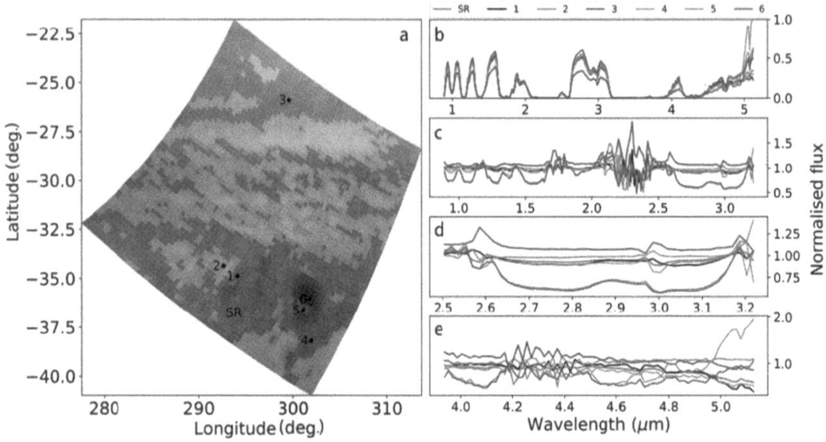

Figure 8.7 Subplot a: Map of Cassini/VIMS cube V1581233933, colored according to the 5 different clusters (SR, 2, 3, 5, 6) identified by the neural network. Subplot b: Atmospheric spectra at locations marked on the map. The SR spectrum and label correspond to the mean spectrum of the blue (mid gray in print version) designated area on the map. Subplots c, d & e: Same as subplot b but with the map's mean spectrum subtracted for clarity.

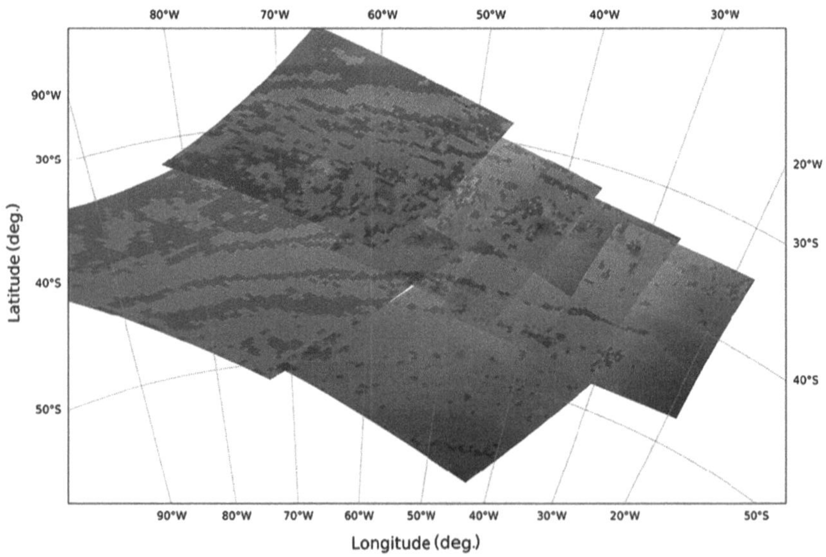

Figure 8.8 Cloud distribution as mapped by the neural network across six overlapping data sets. Colors are identical to Fig. 8.7. It is apparent that the SR feature (blue (mid gray in print version)) occurs in the vicinity of dark storms.

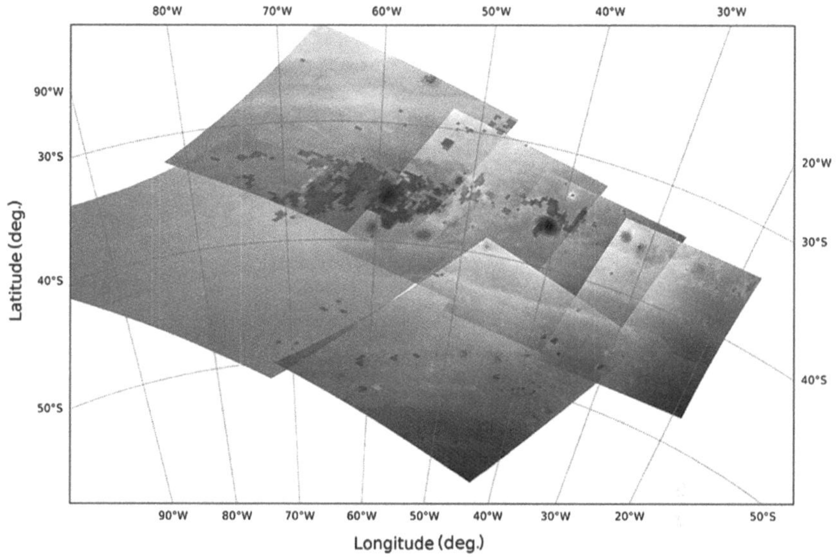

Figure 8.9 Same as Fig. 8.8 but only for the 'stormy region' (SR) referred to in the text and Fig. 8.7.

Each cluster is distinguished by its absorption and scattering characteristics, indicative of the cloud structure and gas composition. Most salient are the spectral differences between the region surrounding the dark storm features (blue (mid gray in print version) region here-forth referred to as SR) in contrast to the unperturbed regions (red/orange (mid gray/light gray in print version)), and the unique signatures of the black storms (purple/green (dark gray/mid gray in print version)). Spectra 1 & 4 in Fig. 8.7 are examples of spectra belong to the SR region. We find that regions unaffected by the storm (e.g. spectra 2 and 3) display the brightest albedo at 1 - 2 μm (Fig. 8.7). At these wavelengths, the bands of the well-mixed CH_4 are modulated by clouds, the brightness of which indicates high aerosols. In contrast, the region surrounding the storm features (blue (mid gray in print version)) is dimmer at 1 - 2 μm, suggesting either lower clouds, or, more likely, given the great extent of the SR region, these are spectrally darker clouds, as postulated by B09. This interpretation indicates that the blue (mid gray in print version) regions encapsulate current and prior storms darkened by the upwelling material of lower albedo.

The brightest of these features inside the extended storm region (spectrum 1 in Fig. 8.7), forms an "S" feature, which coincides with the position of the electric discharges B09 measured on February 9, 2008 by Cassini's

Radio and Plasma Wave Science (RPWS) instrument [30–32]. We find the "S" feature to be the proverbial "tip of the iceberg" of a much larger region of upwelling.

Once trained on a hyperspectral cube, the neural network allows us to quickly and accurately map salient spectral regions over multiple heterogeneous data sets, spanning a large area of the planet. Whereas our analysis focused on the training cube so far, we now use the NN to map a much larger region by including five additional data cubes that encompass the original storm, as well as two smaller storms eastwards. As shown in Figs. 8.8 and 8.9 we detect the presence of the SR feature, significantly beyond the spatial coverage initially reported by B09. In addition, we find similar spatial/spectral signatures of the stormy region around a smaller dark storm to the east (42°W), indicating that areas of upwelling are common around dark storms on Saturn.

References

[1] H.R. Kerner, K.L. Wagstaff, B.D. Bue, P.C. Gray, J.F. Bell, H. Ben Amor, Toward generalized change detection on planetary surfaces with convolutional autoencoders and transfer learning, IEEE Journal of Selected Topics in Applied Earth Observations and Remote Sensing 12 (10) (2019) 3900–3918.

[2] Kiri Wagstaff, You Lu, Alice Stanboli, Kevin Grimes, Thamme Gowda, Jordan Padams, Deep Mars: Cnn classification of Mars imagery for the pds imaging atlas, in: Proceedings of the Thirtieth Annual Conference on Innovative Applications of Artificial Intelligence, 2018, https://www.aaai.org/ocs/index.php/AAAI/AAAI18/paper/view/16040.

[3] V.T. Bickel, C. Lanaras, A. Manconi, S. Loew, U. Mall, Automated detection of lunar rockfalls using a convolutional neural network, IEEE Transactions on Geoscience and Remote Sensing 57 (6) (2019) 3501–3511.

[4] Alistair Francis, Panagiotis Sidiropoulos, Jan-Peter Muller, Cloudfcn: accurate and robust cloud detection for satellite imagery with deep learning, Remote Sensing (ISSN 2072-4292) 11 (19) (2019), https://doi.org/10.3390/rs11192312.

[5] Hannah R. Kerner, Kiri L. Wagstaff, Brian D. Bue, Danika F. Wellington, Samantha Jacob, Paul Horton, James F. Bell, Chiman Kwan, Hani Ben Amor, Comparison of novelty detection methods for multispectral images in rover-based planetary exploration missions, Data Mining and Knowledge Discovery (ISSN 1384-5810) (2020), https://doi.org/10.1007/s10618-020-00697-6.

[6] A.M. Saranathan, M. Parente, Adversarial feature learning for improved mineral mapping in CRISM images, in: Lunar and Planetary Science Conference, March 2019, p. 2698.

[7] Hao Wu, Saurabh Prasad, Convolutional recurrent neural networks for hyperspectral data classification, Remote Sensing (ISSN 2072-4292) 9 (3) (2017), https://doi.org/10.3390/rs9030298.

[8] J. Yang, Y. Zhao, J.C.W. Chan, C. Yi, Hyperspectral image classification using two-channel deep convolutional neural network, in: 2016 IEEE International Geoscience and Remote Sensing Symposium (IGARSS), July 2016, pp. 5079–5082.

[9] X.X. Zhu, D. Tuia, L. Mou, G.S. Xia, L. Zhang, F. Xu, F. Fraundorfer, Deep learning in remote sensing: a comprehensive review and list of resources, IEEE Geoscience and Remote Sensing Magazine (ISSN 2473-2397) 5 (4) (Dec 2017) 8–36, https://doi.org/10.1109/MGRS.2017.2762307.
[10] I.P. Waldmann, C.A. Griffith, Mapping Saturn using deep learning, Nature Astronomy 3 (Apr 2019) 620–625, https://doi.org/10.1038/s41550-019-0753-8.
[11] R.H. Brown, K.H. Baines, G. Bellucci, J.-P. Bibring, B.J. Buratti, F. Capaccioni, P. Cerroni, R.N. Clark, A. Coradini, D.P. Cruikshank, P. Drossart, V. Formisano, R. Jaumann, Y. Langevin, D.L. Matson, T.B. McCord, V. Mennella, E. Miller, R.M. Nelson, P.D. Nicholson, B. Sicardy, C. Sotin, The Cassini visual and infrared mapping spectrometer (Vims) investigation, SSR 115 (December 2004) 111–168, https://doi.org/10.1007/s11214-004-1453-x.
[12] T. Fouchet, T.K. Greathouse, A. Spiga, L.N. Fletcher, S. Guerlet, J. Leconte, G.S. Orton, Stratospheric aftermath of the 2010 Storm on Saturn as observed by the TEXES instrument. I. Temperature structure, Icarus 277 (October 2016) 196–214, https://doi.org/10.1016/j.icarus.2016.04.030.
[13] L.N. Fletcher, B.E. Hesman, P.G.J. Irwin, K.H. Baines, T.W. Momary, A. Sanchez-Lavega, F.M. Flasar, P.L. Read, G.S. Orton, A. Simon-Miller, R. Hueso, G.L. Bjoraker, A. Mamoutkine, T. del Rio-Gaztelurrutia, J.M. Gomez, B. Buratti, R.N. Clark, P.D. Nicholson, C. Sotin, Thermal structure and dynamics of Saturn's northern springtime disturbance, Science 332 (June 2011) 1413, https://doi.org/10.1126/science.1204774.
[14] J.K. Barstow, P.G.J. Irwin, L.N. Fletcher, R.S. Giles, C. Merlet, Probing Saturn's tropospheric cloud with Cassini/VIMS, Icarus 271 (June 2016) 400–417, https://doi.org/10.1016/j.icarus.2016.01.013.
[15] A. Sánchez-Lavega, E. García-Melendo, S. Pérez-Hoyos, R. Hueso, M.H. Wong, A. Simon, J.F. Sanz-Requena, A. Antuñano, N. Barrado-Izagirre, I. Garate-Lopez, J.F. Rojas, T. Del Río-Gaztelurrutia, J.M. Gómez-Forrellad, I. de Pater, L. Li, T. Barry, An enduring rapidly moving storm as a guide to Saturn's Equatorial jet's complex structure, Nature Communications 7 (November 2016) 13262, https://doi.org/10.1038/ncomms13262.
[16] K.H. Baines, M.L. Delitsky, T.W. Momary, R.H. Brown, B.J. Buratti, R.N. Clark, P.D. Nicholson, Storm clouds on Saturn: lightning-induced chemistry and associated materials consistent with Cassini/VIMS spectra, Planetary and Space Science 57 (December 2009) 1650–1658, https://doi.org/10.1016/j.pss.2009.06.025.
[17] Ian Jolliffe, Principal Component Analysis, Springer Verlag, New York, 2002.
[18] Christopher M. Bishop, Pattern Recognition and Machine Learning (Information Science and Statistics), Springer-Verlag New York, Inc., Secaucus, NJ, USA, ISBN 0387310738, 2006.
[19] Caitlin A. Griffith, Paulo F. Penteado, Jake D. Turner, Catherine D. Neish, Giuseppe Mitri, Nicholas J. Montiel, Ashley Schoenfeld, Rosaly M.C. Lopes, A corridor of exposed ice-rich bedrock across Titan's tropical region, Nature Astronomy 3 (Apr 2019) 642–648, https://doi.org/10.1038/s41550-019-0756-5.
[20] Bernhard Schoelkopf, Alexander J. Smola, Klaus-Robert Mueller, Kernel principal component analysis, in: Advances in Kernel Methods, MIT Press, Cambridge, MA, USA, 1999, pp. 327–352.
[21] S.X. Yu, J. Shi, Multiclass spectral clustering, in: Proceedings Ninth IEEE International Conference on Computer Vision, vol. 1, Oct 2003, pp. 313–319.
[22] Martin Ester, Hans-Peter Kriegel, Jörg Sander, Xiaowei Xu, A density-based algorithm for discovering clusters in large spatial databases with noise, in: KDD'96: Proceedings of the Second International Conference on Knowledge Discovery and Data Mining, AAAI Press, 1996, pp. 226–231.
[23] Stella X. Yu, Jianbo Shi, Multiclass spectral clustering, in: Proceedings of the Ninth IEEE International Conference on Computer Vision, vol. 2, ICCV'03, Washington,

DC, USA, IEEE Computer Society, ISBN 0-7695-1950-4, 2003, p. 313, http://dl.acm.org/citation.cfm?id=946247.946658.

[24] Andrew Y. Ng, Michael I. Jordan, Yair Weiss, On spectral clustering: analysis and an algorithm, in: Advances in Neural Information Processing Systems, MIT Press, 2001, pp. 849–856.

[25] F. Pedregosa, G. Varoquaux, A. Gramfort, V. Michel, B. Thirion, O. Grisel, M. Blondel, P. Prettenhofer, R. Weiss, V. Dubourg, J. Vanderplas, A. Passos, D. Cournapeau, M. Brucher, M. Perrot, E. Duchesnay, Scikit-learn: machine learning in Python, Journal of Machine Learning Research 12 (2011) 2825–2830.

[26] Ulrike Luxburg, A tutorial on spectral clustering, Statistics and Computing (ISSN 0960-3174) 17 (4) (December 2007) 395–416, https://doi.org/10.1007/s11222-007-9033-z.

[27] Yoshua Bengio, Learning Deep Architectures for AI, vol. 2, Now Publishers Inc, 2009.

[28] Ian Goodfellow, Yoshua Bengio, Aaron Courville, Deep Learning, MIT Press, 2016, http://www.deeplearningbook.org.

[29] Martín Abadi, Ashish Agarwal, Paul Barham, Eugene Brevdo, Zhifeng Chen, Craig Citro, Greg S. Corrado, Andy Davis, Jeffrey Dean, Matthieu Devin, Sanjay Ghemawat, Ian Goodfellow, Andrew Harp, Geoffrey Irving, Michael Isard, Yangqing Jia, Rafal Jozefowicz, Lukasz Kaiser, Manjunath Kudlur, Josh Levenberg, Dandelion Mané, Rajat Monga, Sherry Moore, Derek Murray, Chris Olah, Mike Schuster, Jonathon Shlens, Benoit Steiner, Ilya Sutskever, Kunal Talwar, Paul Tucker, Vincent Vanhoucke, Vijay Vasudevan, Fernanda Viégas, Oriol Vinyals, Pete Warden, Martin Wattenberg, Martin Wicke, Yuan Yu, Xiaoqiang Zheng, TensorFlow: large-scale machine learning on heterogeneous systems, https://www.tensorflow.org/, 2015, Software available from tensorflow.org.

[30] D.A. Gurnett, W.S. Kurth, G.B. Hospodarsky, A.M. Persoon, T.F. Averkamp, B. Cecconi, A. Lecacheux, P. Zarka, P. Canu, N. Cornilleau-Wehrlin, P. Galopeau, A. Roux, C. Harvey, P. Louarn, R. Bostrom, G. Gustafsson, J.E. Wahlund, M.D. Desch, W.M. Farrell, M.L. Kaiser, K. Goetz, P.J. Kellogg, G. Fischer, H.P. Ladreiter, H. Rucker, H. Alleyne, A. Pedersen, Radio and plasma wave observations at Saturn from Cassini's approach and first orbit, Science 307 (February 2005) 1255–1259, https://doi.org/10.1126/science.1105356.

[31] G. Fischer, W.S. Kurth, U.A. Dyudina, M.L. Kaiser, P. Zarka, A. Lecacheux, A.P. Ingersoll, D.A. Gurnett, Analysis of a giant lightning storm on Saturn, Icarus 190 (October 2007) 528–544, https://doi.org/10.1016/j.icarus.2007.04.002.

[32] Georg Fischer, Donald A. Gurnett, William S. Kurth, Ferzan Akalin, Philippe Zarka, Ulyana A. Dyudina, William M. Farrell, Michael L. Kaiser, Atmospheric electricity at Saturn, SSR 137 (June 2008) 271–285, https://doi.org/10.1007/s11214-008-9370-z.

[33] Nitish Srivastava, Geoffrey Hinton, Alex Krizhevsky, Ilya Sutskever, Ruslan Salakhutdinov, Dropout: a simple way to prevent neural networks from overfitting, Journal of Machine Learning Research (ISSN 1532-4435) 15 (1) (January 2014) 1929–1958, https://www.cs.toronto.edu/~rsalakhu/papers/srivastava14a.pdf.

CHAPTER 9

Machine learning for planetary rovers☆

Masahiro Ono[a], Brandon Rothrock[a], Yumi Iwashita[a], Shoya Higa[a], Virisha Timmaraju[a], Sami Sahnoune[a], Dicong Qiu[b,a], Tanvir Islam[a], Annie Didier[a], Christopher Laporte[a], Deegan Atha[a], Vivian Sun[a], Kyohei Otsu[a], Mike Paton[a], Olivier Lamarre[c,a], Shreyansh Daftry[a], R. Michael Swan[a], Adam Stambouli[a], Flynn Chen[d,a], Bhavin Shah[a], Kathryn Stack[a], and Chris Mattman[a]

[a]Jet Propulsion Laboratory, California Institute of Technology, Pasadena, CA, United States
[b]Carnegie Mellon University, Pittsburgh, PA, United States
[c]University of Toronto, Toronto, ON, Canada
[d]Yale University, New Haven, CT, United States

9.1 Introduction

There is an urgent need for significantly enhancing on-board autonomy of future rover missions. For example, an ambitious Lunar rover mission concept called *Intrepid* (Fig. 9.1-Left) aims to drive ~1800 km in 4 years to cover geological records spanning over four billion years [2]. A typical operation of Intrepid would involve a ~20 km traverse in ~50 hours to drive to the next scientific region of interest. While the typical autonomous driving distance of NASA's latest Mars rover, *Perseverance* (Fig. 9.1-Right), is up to hundreds of meters per sol (Martian day) at the speed of 4.2 cm/sec or ~150 m/hr, the Intrepid rover would need to extend the mean distance between human intervention to 6 - 16 km. For another example, the high-priority mission elements identified by a report from the Mars Architecture Strategy Working Group include (1) affordable, mobile geological explorers for characterizing diverse habitable environments, (2) in situ mid-latitude ice sampling, and (3) in situ polar layer deposit climate record determination, which could use mobility to examine exposed layers over a multi-kilometer traverse [1]. A study at JPL suggests that Mars rover range >100 km per Earth year may be possible and that this would enable

☆ Government sponsorship acknowledged. Contribution prepared by the Contributor on behalf of JPL/Caltech.

Machine Learning for Planetary Science
https://doi.org/10.1016/B978-0-12-818721-0.00019-7
Copyright © 2022 Elsevier Inc.
All rights reserved.

Figure 9.1 Left: Artist concepts of the Intrepid Rover (NASA/JPL-Caltech); Right: NASA's Perseverance rover (NASA/JPL-Caltech/MSSS).

visiting multiple geologic type localities in a single mission. Faster driving would require a radical departure from the current paradigm of planetary rover operation that heavily relies on ground-in-the-loop processes, which in turn necessitates more reliable and "human-like" autonomous driving capabilities.

Faster driving would also generate data at an increased rate, while the capacity of Mars-Earth communication remains limited by the laws of physics as well as the availability of relay orbiters and the Deep Space Network. For example, when Perseverance drives with visual odometry, a pair of monocolor navigation camera images is taken every 1 m and stored onboard. This produces ~135 Mbytes of data over a 100-meter drive after downsampling. However, only a subsampled set of the navigation images are downlinked due to the bandwidth limitation in interplanetary communication, and the unsent images are eventually erased due to the limitation in onboard storage, even though Perseverance drives only up to a few hundred meters per Sol. Modern data-intensive instruments, such as hyperspectral imagers and ground-penetrating radars, can easily produce orders of magnitude greater volume than the grayscale navigation images. With such unbalanced data production and transmission rates, the rover must by necessity discard a large amount of data, resulting in potentially lost science opportunities. In fact, there have been a number of serendipitous discoveries made by existing and past Mars rovers, which would have been missed in a high-speed driving mission [3]. For example, the boxwork features unexpectedly found at Matijevic Hill by Opportunity (Fig. 9.2-Left) turned out to be an evidence of neutral water rather than acidic [4]. The Iron Meteorite named

Figure 9.2 Left: Evidence of neutral water (boxwork features) in Matijevic Hill, discovered by Opportunity. Right: Iron Meteorite "Lebanon" found by Curiosity. These are examples of serendipitous discoveries, which might have been missed by a high-speed rover without on-board interpretation of the scene. Courtesy NASA/JPL-Caltech.

"Lebanon" by Curiosity on Sol 640 is another example of serendipitous discoveries. A fast-driving rover could pass by such scientific features without being noticed by ground operations because image acquisition is not commanded at the moment of encounter or the images are simply not downlinked due to the limitation in the communication capacity.

In summary, future rover missions would be required to drive over a substantially longer distance without missing science opportunities, which in turn demands an elevated level of onboard autonomy and analytical capabilities. To this end, JPL's Machine learning-based Analytics for Rover Systems (MAARS) project developed capabilities to employ modern machine learning methods for vision-based traversability assessment for safe navigation and scientific scene interpretation. In particular, the project focused on the following two capabilities.

1) Risk- and resource-aware AutoNav

The traversability assessment of the conventional autonomous driving capability of Mars rovers, called AutoNav [5], solely depends on stereo vision. For example, Fig. 9.3 shows a navigation camera (Navcam) image taken by the *Opportunity* rover, as well as the terrain topography obtained from the onboard images through stereo vision. Note that the distinction between two major terrain classes (bedrock and sand) is absent from the topographical representation. Unlike AutoNav, human rover drivers use semantic information in raw images to assess various factors of surface traversability, such as slip, sinkage, and driving energy. Likewise, onboard interpretation of semantic information would enable the rover to incorporate the additional

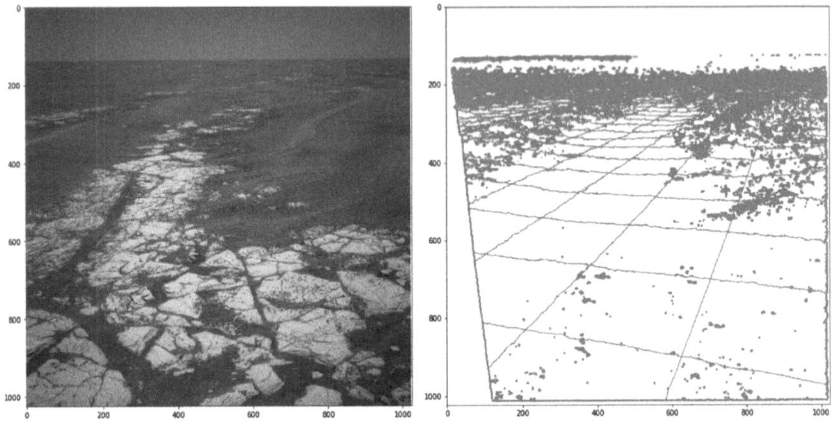

Figure 9.3 A navigation camera image captured by the *Opportunity* rover on its Sol 809. Left: raw image, right: terrain mesh created from the same image. Courtesy NASA/JPL-Caltech.

risk factors of terrain type, rock hazards, and wheel slip into consideration when autonomously planning a path. Combined with on-board data interpretation using machine learning, this would also enable future rovers to plan a long-range path with diverse objectives, such as science gain maximization, while increasing the level of safety. We collectively call these capabilities the risk- and resource-aware AutoNav, which will be detailed in Section 9.2.

Drive-by science (DBS)

As we mentioned above, rovers take navigation images every meter or so while driving autonomously, which are often not fully downlinked. If an onboard algorithm can analyze these images to look for features of scientific interest, rovers would be able to avoid missing serendipitous encounters. Hence, we envision the *drive-by science (DBS)* capability, which analyzes data, in particular the images acquired while driving, to enable scientists on the ground to detect interesting science features and selectively downlink relevant data without interrupting drives.

Our approach to realizing DBS is multifold. We use machine learning to extract a compact representation of collected data, such as feature vectors or image captions, which can be used as a summary of the data for downlink. Alternatively, the ground scientists can send a query to the onboard database of the rover to specify the kind of data they are seeking for by keywords or sample images (i.e., image similarity search). Furthermore, it

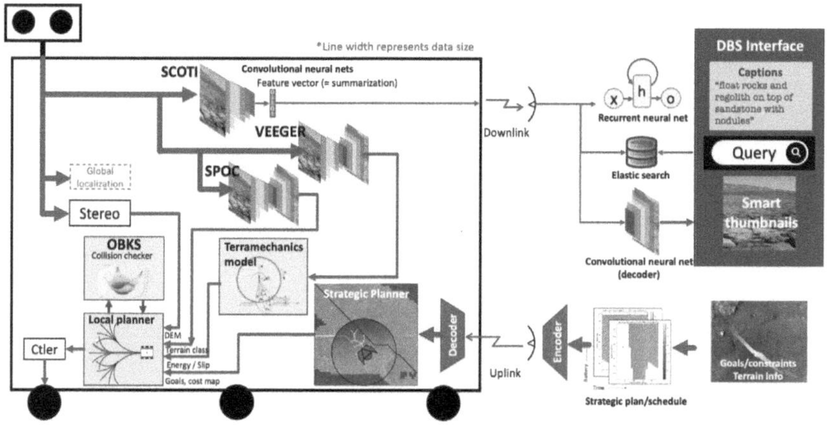

Figure 9.4 The overall composition of the MAARS system, consisting of multiple algorithmic components described in Sections 9.2 and 9.3.

would allow the rover to autonomously plan a path, select activities, and target the instruments to achieve prescribed science goals with a significantly fewer number of ground-in-the-loop cycles. We will explain the DBS capability in detail in Section 9.3.

Both capabilities consist of multiple algorithmic components. Fig. 9.4 shows the overall composition of the MAARS system. In the reminder of this chapter, Sections 9.2 and 9.3 will provide overviews of the Risk- and Resource-Aware AutoNav and Drive-by Science capabilities, respectively, as well as high-level descriptions of their algorithmic components. Then, Section 9.5 will conclude the chapter by presenting the ongoing and future work.

9.2 Risk- and resource-aware AutoNav

9.2.1 Overview

The local path planning capability of MAARS is provided by a combination of several new and existing algorithms. On the perception side, the Soil Property and Object Classification (SPOC) terrain classifier [6] leverages a deep-learning-based segmentation network to predict terrain type, and in particular identify sand hazards that cannot be identified geometrically through stereo vision. Similarly, while large rocks can be identified and avoided from stereo, small embedded rocks that can cause wheel damage are difficult to identify, as well as larger rock hazards at further range

Figure 9.5 Sand trap has been a major risk to Mars rovers. Left: Opportunity at Purgatory Dune on Sol 461 (May 11, 2005); Right: Curiosity at Hidden Valley on Sol 712 (Aug 7, 2014). Courtesy NASA/JPL-Caltech.

where stereo vision cannot resolve them. To address this, a rock detection approach is proposed that fuses both geometric stereo and monocular model-based semantic vision. Finally, we present a driving energy prediction algorithm called *VeeGer* [22] that combines terrain classification and obstacle detection. The resource awareness comes from VeeGer, where the estimated energy usage is used as a part of the cost function of the path planner.

On the path planning side, a standard tree planner is combined with collision checking. Users can choose from three collision checking algorithms: ACE (approximate kinematic settling) [8], which is used by the Mars 2020 rover with a collision checking algorithm, the probabilistic extension of ACE called p-ACE [9], and optimization-based kinematic settling (OBKS).

9.2.2 Terrain classification

Terrain has been a major source of risk for Mars rovers. For example, Mars Rover Spirit ended its mission after becoming embedded in sand in 2009. *Opportunity* got stuck in a location on the traverse called "Purgatory Dune" and took six weeks to escape (Fig. 9.5-Left). Curiosity experienced excessive slip on a sandy surface in "Hidden Valley" at an average slope of merely 4 degrees (Fig. 9.5-Right). Nevertheless, current Mars rovers lack the onboard capability to identify nongeometric hazards like sand.

Obtaining a slip prediction model for Mars rovers is a challenging task, however. In general, there are two approaches for slip prediction. One is a physics-based approach, where the mechanical forces and torques between the wheels and a soil are computed using terramechanics. However, this approach requires identifying soil parameters such as cohesion and friction angle [10], which is highly challenging. This approach will be presented in Section 9.2.4. The other is an empirical approach, where the relation between slip, slope angle, and terrain type is estimated through regression based on past driving data. However, this approach cannot be used at the early phase of the surface mission due to the lack of sufficient driving data. Data from past rover missions, such as Mars Exploration Rovers (MER), is not usable due to the significant difference in the size and mass of the rovers as well as the variation in soil properties between landing sites.

As a result, slip prediction for the Curiosity rover has been primarily made by a model obtained from experiments on Earth [11]. The model consists of three functions, corresponding to three distinct terrain types: sand, cohesive soil, and bedrock. For each terrain type, the function gives the estimate of slip ratio as a function of the slope angle. A substantial volume of driving data is available from past and ongoing rover missions. For example, the first four years of the Curiosity rover operation recorded ~8000 instances of slip, detected by the difference between the wheel odometry and the visual odometry. However, the regression requires knowledge of terrain type, which can only be identified from images. Manually identifying terrain type for thousands of samples is not practical.

We developed the Soil Property and Object Classification model to enable automatic terrain class labeling from rover's navigation camera (Navcam) images, which subsequently enabled the generation of an empirical slip prediction model from in-situ mobility data.

The SPOC model architecture is derived from "DeepLab" [12], which is a fully-convolutional segmentation model. The network front end of an early version of SPOC utilized a VGG architecture [13] modified to use dilated convolutions [14]. These specialized convolutions effectively increase the receptive field of the filters without increasing the filter size. The latest version of SPOC uses DeepLabv3+ with a ResNet-101 backend, pretrained on ImageNet [15–17]. The network is trained with standard backpropagation and stochastic gradient descent, and takes on the order of 6 hours to train on a modern desktop GPU. The latest model was trained by ~160K labels annotated by citizen scientists [19] and tested against ~1K

Table 9.1 Confusion matrix of a SPOC model trained with ~160K labels annotated by citizen scientists. The values in the matrix are calculated with respect to the 3 label agreement test set, created by JPL experts. Overall accuracy is 96.67%.

		Predicted			
		Soil	**Bedrock**	**Sand**	**Big Rock**
Actual	Soil	99.10	0.32	0.57	0.01
	Bedrock	3.64	94.90	0.37	1.09
	Sand	0.88	5.62	93.45	0.05
	Big Rock	6.76	0	0	93.24

expert labels annoted by rover planners and scientists. The overall pixel accuracy was 96.67% (weighted average over the classes). Table 9.1 shows the confusion matrix. Fig. 9.6 shows sample results on Navcam images from Curiosity. Inference of a surface image can be processed entirely within the GPU in under 200 ms. The processing time of a 513x513 image on Jetson's TX2, which is comparable to future high-performance spacecraft computing [18], was less than 1 sec along with other MAARS algorithms running in parallel. More details on SPOC and additional results are reported in [6,7].

9.2.3 Rock hazard detection

One of the challenges with stereo-based vision is the inability to resolve small obstacles such as sharp embedded rocks that can incur wheel damage [20]. Resolving these structures becomes progressively more difficult with increasing distance from the rover. Our approach to this problem combines the use of geometric stereo vision with a monocular algorithm that uses a convolutional instance-based segmentation model. The overall system is designed to semantically identify hazardous rocks in 3D, as well as identify their size in terms of height from the local ground plane. This is first studied from the perspective of automation within rover operations to support improved path planning on the ground, but the ultimate goal is to transition this capability on-board the rover.

The monocular rock hazard detection algorithm is based on Mask-RCNN [21]. The algorithm detects unique rocks and segments their boundaries. There are two modes that this detector needs to run in. The first utilizes Navcam images, which captures the near-field for navigation and are grayscale for MSL. The second utilizes Mastcam images, which captures the far-field for primarily science purposes and are in RGB color.

Figure 9.6 Representative results of the SPOC terrain classifier on navigation camera (Navcam) images from Perseverance. Courtesy NASA/JPL-Caltech.

Both modes utilize the stereo data from the Navcam system as input. To train this system, rocks from a collection of historical Navcam and Mastcam images from MSL were manually annotated. Since the abundance, size, and scale variations of the rocks make it difficult to completely label all rocks in the images, we modified the loss function for Mask-RCNN so that any unlabeled region would be considered unknown and not backpropagated. Therefore, only regions where the background is explicitly labeled will in-

Figure 9.7 (left) Original Navcam Images. (right) Rock Detection Overlays. Courtesy NASA/JPL-Caltech.

fluence the weights in the network. During training, if a region proposal overlaps with the background region beyond a threshold it was considered background. If a proposed region was in the unknown region, it was ignored during backpropagation.

For the Navcam detector, rocks are detected and segmented using a single monocular grayscale image. The size of the rock is then estimated by using the stereo data within the segmented region. To find the height of the region, all of the stereo points are rotated so that the slope of the surface region is zero. After the rotation the difference between the maximum and minimum height values within a cylinder around the region determines the height. A sample of the resulting detections on Navcam images is in Fig. 9.7.

For the Mastcam detector, dense stereo is uncommon so a different method of height estimation is required. For this algorithm, the segmented rock is searched for the maximum difference in pixel height values. This pixel height can be converted to a world frame height through camera transformations with a known range, which is found by searching through sparse stereo hits within the segmented region and surrounding area. Examples of Mastcam detections are shown in Fig. 9.8.

9.2.4 Vision-based slip and driving energy prediction

While exiting path planners for Mars rovers, such as AutoNav and its successor, Enhanced AutoNav, simply minimizes the path length or predicted

Figure 9.8 (left) Original Mastcam Image. (right) Rock Detection Overlay. Courtesy NASA/JPL-Caltech.

drive time to the goal, human rover planners consider a richer set of metrics to plan a path. In particular, predicted slip is an important metric as it results in decreased localization accuracy at best and a loss of vehicle at worst. If the rover is solar-powered, driving energy is another important consideration.

On-board processing of the semantic information in images enables rovers to plan a path in a more human-like way, considering slip and driving energy. To this end, we developed a vision-based algorithm to autonomously predict the slip and driving energy consumption using machine learning [22]. We developed and compared two machine-learning models: VeeGer-EnergyNet and VeeGer-TerramechanicsNet. The former is an end-to-end black-box model that directly predict the driving energy from input images, while the latter is a semimodel-based approach where the ML mode estimates intermediate parameters describing the terrain properties, which are fed to a simplified terramechanics model to compute the driving energy. The prediction of slip is derived as a byproduct. Both models can be trained with self-supervised learning.

Both models take an RGB image and a range image, as shown in Fig. 9.9-(b), as inputs. We use a modified PNASNet-5 [23] to combine the two inputs. Fig. 9.9 shows an example result of predicting power by Veeger-TerramechanicsNet. The training and test data was collected in JPL's Mars Yard, using the six-wheeled Athena test rover. A comparison

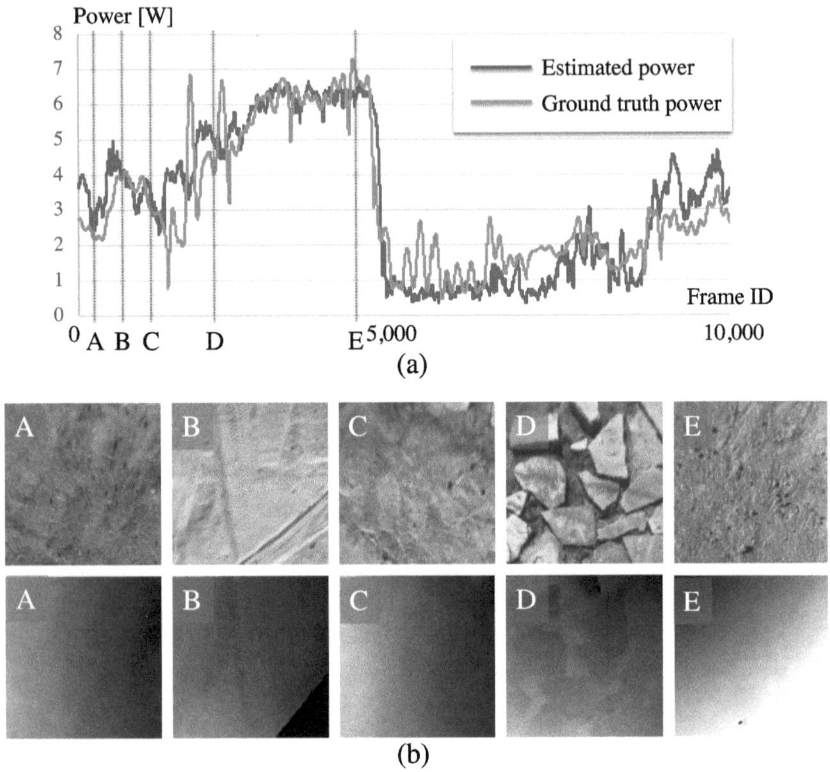

Figure 9.9 (a) Energy consumption estimated from terramechanics parameters predicted with Veeger-TerramechanicsNet, (b) RGB images (top raw) and depth images (bottom) at time A-E in (a). Courtesy NASA/JPL-Caltech.

of the two approaches shows that Veeger-TerramechanicsNet outperforms VeeGer-EnergyNet. Details of the experiment are described in [22].

In VeeGer-TerramechanicsNet, the ML model takes images as inputs and predicts intermediate variables, which involve terramechanics parameters specifying the property of the terrain, and a simplified terramechanics model maps the terramechanics parameters and the observed slope to a prediction of drive energy. To create a training data set, we solve the inverse problem to predict the terramechanics parameters from drive data collected in the Mars Yard, hence no manual labeling was needed. The inverse problem was formulated as the following constrained optimization, where the wheel load is calculated based on the simplified terramechanics model and estimated wheel load balance:

$$
\begin{aligned}
\min_{k_{normal}, n, k_{shear}, z_s, z_d} \quad & \left[(1 - F_z/W)^2 + (1 - M_y/T)^2\right] \\
\text{s.t.} \quad & 0.0 \leq k_{normal} \leq 9.0e^5 \\
& 0.1 \leq n \leq 1.0 \\
& 0.0 \leq k_{shear} \leq 1.0 \\
& 1e^{-4} \leq z_s \leq 2(r + h)/3 \\
& 0.0 \leq z_d \leq 0.0015
\end{aligned}
\tag{9.1}
$$

where W, T, F_z, and M_y are an estimated wheel load, measured wheel torque, calculated wheel load, and calculate wheel torque based on the simplified terramechanics model, respectively. Moreover, k_{normal}, n, k_{shear}, z_s, z_d are simplified terramechanics parameters used to calculate F_z and M_y. More details on VeeGer are described in [22].

9.3 Drive-by-science

9.3.1 Overview

The objective of drive-by-science is to autonomously downselect a subset of data, such as images, collected while the rover is driving such that scientifically valuable features are not missed. The volume of downselected images is limited by the rover's downlink bandwidth. The notion of science value is of course difficult to objectively define; our approach characterizes this as a combined relevance score produced by a trained model to select images according to a specification provided by mission scientists for a given planning period. This specification is provided in natural language using relevant geologic vocabulary. We present a model called SCOTI (Scientific Captioning Of Terrain Images) that automatically generates captions of Mars surface imagery, trained on expert annotations from Mars geologists. We also incorporate the ability to specify image similarity using a distance measure on extracted image features, which can also be used as a measure of image diversity. The captioning and similarity models are packaged into a content-based search system that is the primary interface for how scientists will interact with the rover. Scientists can provide a set of text queries, similarity examples, diversity criteria, filters, and more for what imagery they expect to be of scientific interest for the next planning cycle. This specification is then uploaded to the rover, which autonomously downselects a set of images taken during a given drive.

Figure 9.10 Sample outputs from SCOTI, which generates a natural language sentence describing a given image. Courtesy NASA/JPL-Caltech.

9.3.2 SCOTI: scientific captioning of terrain images

SCOTI takes images as an input, and outputs an English sentence explaining the geological content of the image. It is built upon the "Show, Attend, and Tell" model [24], which is an attention-based model for image caption generation. The model is trained using standard backpropagation techniques with stochastic gradient descent using adaptive learning rate algorithms (RMSProp [25] and Adam [26]). The encoder part of the model uses a convolutional neural network for extracting a set of feature vectors from the image, also known as annotation vectors. The decoder part of the model uses a long short-term memory (LSTM) network that produces a caption by generating the words sequentially conditioned on the context vector as well as the previous hidden state. The VGG19 [13] pretrained on ImageNet [17] is used as the feature extractor. We used BLEU scores [27] to measure the similarity between the expert captions and predicted captions. These scores were used as a metric to optimize our regularization strategy and model architecture. The LSTM network is trained from scratch with 1000 image captions created by expert geologists. It achieved a BLEU-4 score of 0.85 on validation sets. Sample outputs from SCOTI are shown in Fig. 9.10.

SCOTI enables the *"interplanetary Google image search"* concept (Fig. 9.11). Like search engines on the internet, the users (i.e., ground scientists) do not need to download all the data for analysis; instead, they express their interest in natural language words, which are uplinked to the rover, and only

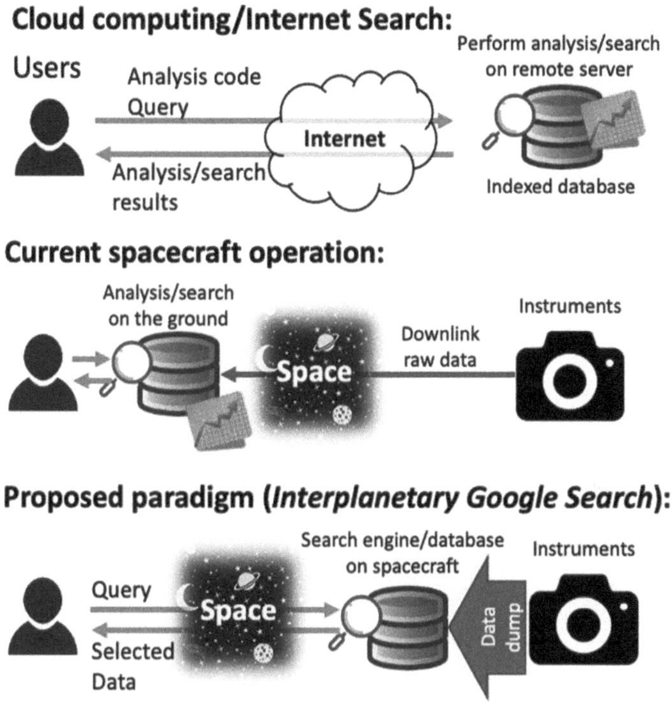

Figure 9.11 Concept of "Interplanetary Google Search".

relevant data is returned to the Earth. Captions from SCOTI in the on-board database are compared against the query words for data prioritization. Alternatively, SCOTI can be used as a data summarization capability, where the ground scientists use the downlinked captions to decide which images should be downlinked at full resolution. There is a ground application, too: currently, scientists needs to manually inspect tens of thousands of rover images to find images with specific geological features such as veins, nodules, crossbedded layers and more. SCOTI allows them to find images of interest through a text-based search similar to the ones on the Internet.

9.3.3 Image similarity search

We also developed an image similarity search capability, which returns a list of images that are semantically similar to a query image provided by users. This ability is enabled by extracting an intermediate layer from the VGG19 image encoding portion of SCOTI (Conv5_3) and compare the feature vector against the feature vector of other images using the cosine distance metric to find the most similar or dissimilar (in the case of novelty search) images.

Like SCOTI, this capability could be used for on-board data prioritization and ground-based data search. For the on-board application, the query image is first converted into a feature vector, which is much smaller in terms of data size. The uplinked feature vector is compared against the feature vectors of on-board images to prioritize them. The ground-based similarity search provides an additional way of quickly finding images of interest in image datasets.

9.3.4 DBS interface

To enable any scientist or rover operator to interact with all of these DBS capabilities, we prototyped a web application (Fig. 9.12) that implements a Google Maps-like search interface for the Martian surface explored by MSL. To start with, a user may perform a text search on the SCOTI-generated image captions after which the paginated side-bar displays the query results by showing the picture, the generated caption, and other relevant metadata. Additionally, the geographic locations of results on the current page are marked by pins on orbital satellite imagery, while the geographic distribution of all results is visualized using a dynamic heat map. If the user is interested in specific geographic regions, a drawing tool allows them to specify areas of interest and only search and see results within those regions. We also integrated the image similarly search capability. Once an image is selected, a simple click of a button allows a user to see similar images displayed in the same way as the text query results. Ultimately, making these capabilities available through a centralized user-friendly application greatly helps scientists make the best use of the vast trove of Martian images to make quicker and better-informed decisions.

9.3.5 DBS experiment with scientists

To test our Drive-By-Science (DBS) capabilities, we conducted a "mock science mission," in which four scientists were asked to perform scientific tasks using the DBS capability. We put the subjects in a scenario simulating a conceptual rover mission, where the rover collects significantly more images that can be sent. The subjects used DBS tools to downselect the images in a way to maximize the accuracy of scientific interpretation in the given tasks.

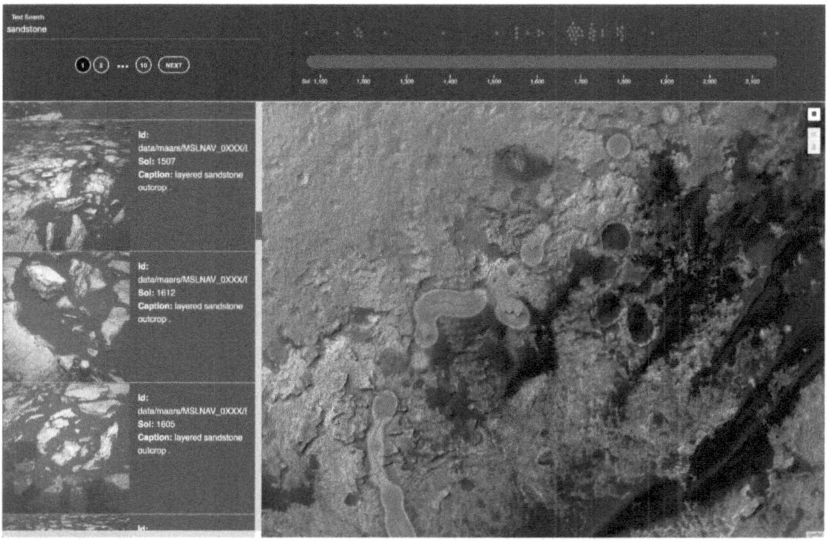

Figure 9.12 The prototyped DBS Interface, which allows scientists to perform search on rover images by keywords or image similarity. Courtesy NASA/JPL-Caltech.

9.3.5.1 Methods

The experiment was performed in the following steps. We prepared a dataset, which contained ~ 100 MSL Navcam images sampled from a range of several sols. It was assumed that only 16 images out of the full dataset can be downlinked due to data capacity.

1. First, each subject was asked to select 16 images to downlink in one of the three following ways: i) random selection (the subject has no choice), ii) selection by on-board DBS, where the subject uplinks a set of keywords and the DBS algorithm returns the 16 most relevant images, and iii) ground-based selection with DBS, where the subject selects 16 images based on the auto-generated captions and thumbnails. The assumption made here is that although the rover cannot send back all its raw images, captions and thumbnails of all images are of sufficient size to be downlinked.
2. Next, each subject was asked to perform a scientific task given in a form of a questionnaire using the 16 downlinked images.
3. Then, the subjects were provided with the full dataset.
4. Finally, each subject was asked to perform the same scientific task on the full dataset.

We observed the change in the scientific interpretation before and after seeing the full dataset. If the change is small, that indicates that the downsampled dataset properly represents the full dataset. We repeated this experiment three times with different datasets.

9.3.5.2 Tasks

We developed a questionnaire that was cognizant of the fact that different scientists have different research focuses and may therefore focus on different features. Scientists may interpret the same data differently depending on if they specialize in this particular field of expertise, and differences in interpretation are compounded by the limitations in the image dataset (i.e., there is more uncertainty in interpretation when images are limited in spatial extent and are of relatively coarse resolution). We therefore asked a set of straightforward science interpretation questions to gauge how the subjects would interpret the depositional and diagenetic history of rocks presented in the same image sets, in addition to a second set of questions asking which test scenario they found to be most scientifically useful and why.

9.3.5.3 Results

Due to the small sample size and variability in subjects, sol ranges, and image subsets returned by different scenarios, it was difficult to quantify the performance and accuracy of interpretation in each scenario. For example, subjects had different interpretations when viewing the full dataset, compared to the limited dataset, for each scenario, demonstrating differences in their research background. Subjects who were viewing the same sol range of images but testing different scenarios would also interpret the images differently, but this difference is expected because the different scenarios returned a different subset of the images in that sol range. Fig. 9.13 is the summary of a semiquantifiable questions collected in the experiment. Again, no definitive conclusions can be drawn from this experiment due to its small sample size. While the experiment was meaningful in that we could validate the tools and experimental approach, an obvious next step is to scale up the experiment and collect a statistically significant number of samples. When asked which scenario they preferred, most subjects chose Scenario 4 (the option to select their own images to downlink), which likely represent the preference of scientists to keep the data selection and analysis process under their control. This result suggests that onboard data triage should be performed in a less intrusive way that leaves as much control as possible to the ground scientists.

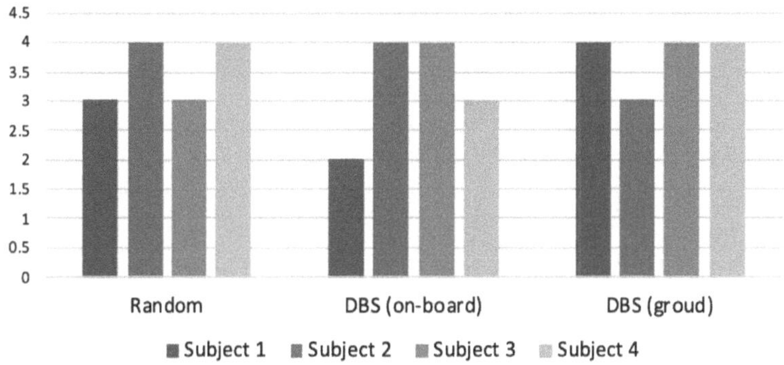

Figure 9.13 Answers from the test subjects to a question: "Do you think the limited dataset was a fair representation of the full dataset, with respect to enabling you to answer the questions re: primary depositional environment and secondary processes?"

9.4 Demonstration on a test rover

Finally, to demonstrate the capabilities presented above, we performed a series of tests at an analog site using a test rover. We have deployed the algorithms presented in this chapter on JPL's Athena test rover, which has two Jetson TX2's, stereo cameras, and six wheels on the rocker-bogie suspension system used by all the existing Mars rovers. The software components were implemented in C++ and Python and integrated with Robot Operation System (ROS). The tests were performed in Arroyo Seco, a river bed right next to JPL, which is dry for most of the seasons. Fig. 9.15 shows the Athena rover in the test area in Arroyo Seco. The obstacles for the rover include large rocks, vegetation, debris, and animal wastes, which are identified by the SPOC terrain classifier trained by labeled images of the test site.

The tests were performed from August to September 2020. The test site in Arroyo served as "Mars" while the operators on "Earth" were located remotely and interacted with the rover over the internet. The Athena rover drove ~500 meters fully autonomously at the speed of ~10 cm/sec using the SPOC terrain classifier (Section 9.2.2) at 2 Hz and the drive cost assessment by VeeGer (Section 9.2.4) at 1 Hz, combined with stereo-based obstacle detection, as shown in Fig. 9.14. All the navigation images were processed in real time by SCOTI (Section 9.3.2) at 0.5 Hz and the feature

Figure 9.14 A screen shot of SPOC and VeeGer visualizations from the rover test at Arroyo Seco in September, 2020.

extraction for image similarity search at 2 Hz. The images, image captions from SCOTI, and the extracted features were sent to a database on an Amazon Web Services (AWS) server every 5 seconds, which simulates the onboard database. The operators at a remote location used the DBS interface (Section 9.3.4) to send queries and extracted images of their interest in real time. All the algorithms ran onboard at the prescribed rate and the rover successfully completed the autonomous drive.

9.5 Conclusion and future work

We have developed, prototyped, and demonstrated a number of promising approaches to utilize machine learning, in particular deep neural networks, to enhance the safety and productivity of rovers. Of particular importance among a variety of potential applications is the semantic interpretation of onboard data, such as images, for providing key information on traversability assessment and scientific feature detection. Such applications would not be limited to rover missions but relevant to a broader range of planetary exploration, including lander, orbiter, and fly-by missions. There are ongoing efforts to infuse these capabilities to flight missions. In particular, the SPOC terrain classifier has been integrated with the ground operation system of Curiosity, and we are undergoing an ongoing effort to upgrade it, based on the latest results from the MAARS project.

Onboard deployment of the capabilities presented in the chapter would require substantially more powerful onboard processors than RAD750, the main CPU of the Curiosity and Perseverance rovers. There is an ongoing joint effort by NASA and AFRL to develop a high-performance spacecraft computing (HPSC) [18]. Meanwhile, the Mars Helicopter Ingenuity, the first powered aerial vehicle on Mars, uses Qualcomm's Snapdragon, a commercial-off-the-shelf SoC (system on chip) [28]. These processors are enablers for potential future onboard deployment of MAARS capabilities.

The capabilities we presented in this chapter would only be a small fraction of the promising potentials of machine learning for planetary exploration that are yet to be explored. Like on Earth, machine learning would soon revolutionize the way we explore the vastly unknown extraterrestrial worlds.

The research was carried out at the Jet Propulsion Laboratory, California Institute of Technology, under a contract with the National Aeronautics and Space Administration (80NM0018D0004).

Figure 9.15 The Athena test rover used for the field tests of capabilities described in this chapter. The test area is located in Arroyo Seco, a river bed next to JPL. Courtesy NASA/JPL-Caltech.

References

[1] R. Jakosky, et al., Mars, The Nearest Habitable World - A Comprehensive Program for Future Mars Exploration; Mars Architecture Strategy Working Group (MASWG) Preliminary Results, https://mepag.jpl.nasa.gov/meeting/2020-06/MASWG-preliminary-results-MEPAG-26June2020.pdf, 2020.

[2] M. Robinson, J. Elliott, et al., Intrepid planetary mission concept study report, https://science.nasa.gov/science-pink/s3fs-public/atoms/files/Lunar%20INTREPID.pdf, 2020.

[3] M. Ono, B. Rothrock, C. Mattmann, T. Islam, A. Didier, V.Z. Sun, D. Qiu, P. Ramirez, K. Grimes, G. Hedrick, Make planetary images searchable: content-based search for pds and on-board datasets, in: 50th Lunar and Planetary Science Conference 2019, 2019.

[4] R.E. Arvidson, S.W. Squyres, J.F. Bell, J.G. Catalano, B.C. Clark, L.S. Crumpler, P.A. de Souza, A.G. Fairén, W.H. Farrand, V.K. Fox, R. Gellert, A. Ghosh, M.P.

Golombek, J.P. Grotzinger, E.A. Guinness, K.E. Herkenhoff, B.L. Jolliff, A.H. Knoll, R. Li, S.M. McLennan, D.W. Ming, D.W. Mittlefehldt, J.M. Moore, R.V. Morris, S.L. Murchie, T.J. Parker, G. Paulsen, J.W. Rice, S.W. Ruff, M.D. Smith, M.J. Wolff, Ancient aqueous environments at Endeavour crater, Mars, Science 343 (6169) (2014), https://doi.org/10.1126/science.1248097.

[5] S.B. Goldberg, M.W. Maimone, L. Matthies, Stereo vision and rover navigation software for planetary exploration, in: Proceedings, IEEE Aerospace Conference, vol. 5, IEEE, 2002, p. 5.

[6] B. Rothrock, R. Kennedy, C. Cunningham, J. Papon, M. Heverly, M. Ono, Spoc: deep learning-based terrain classification for Mars rover missions, in: AIAA SPACE 2016, 2016, pp. 1–12.

[7] D. Atha, R.M. Swan, A. Didier, Z. Hasnain, M. Ono, Multi-mission terrain classifier for safe rover navigation and automated science, in: Proceedings, IEEE Aerospace Conference, IEEE, 2022, p. 5.

[8] K. Otsu, G. Matheron, S. Ghosh, O. Toupet, M. Ono, Fast approximate clearance evaluation for rovers with articulated suspension systems, Journal of Field Robotics 37 (5) (2020) 768–785, https://doi.org/10.1002/rob.21892.

[9] S. Ghosh, K. Otsu, M. Ono, Probabilistic kinematic state estimation for motion planning of planetary rovers, in: 2018 IEEE/RSJ International Conference on Intelligent Robots and Systems (IROS), IEEE, 2018, pp. 5148–5154.

[10] F. Zhou, R.E. Arvidson, K. Bennett, B. Trease, R. Lindemann, P. Bellutta, K. Iagnemma, C. Senatore, Simulations of Mars rover traverses, Journal of Field Robotics 31 (1) (2014) 141–160, https://doi.org/10.1002/rob.

[11] M. Heverly, J. Matthews, J. Lin, D. Fuller, M. Maimone, J. Biesiadecki, J. Leichty, Traverse performance characterization for the Mars science laboratory rover, Journal of Field Robotics 30 (6) (2013) 835–846.

[12] L.-C. Chen, G. Papandreou, I. Kokkinos, K. Murphy, A.L. Yuille, Semantic image segmentation with deep convolutional nets and fully connected crfs, in: International Conference on Learning Representations (ICLR), 2015.

[13] K. Simonyan, A. Zisserman, Very deep convolutional networks for large-scale image recognition, arXiv preprint, arXiv:1409.1556.

[14] F. Yu, V. Koltun, Multi-scale context aggregation by dilated convolutions, arXiv preprint, arXiv:1511.07122.

[15] L.-C. Chen, Y. Zhu, G. Papandreou, F. Schroff, H. Adam, Encoder-decoder with atrous separable convolution for semantic image segmentation, in: ECCV, 2018.

[16] K. He, X. Zhang, S. Ren, J. Sun, Deep residual learning for image recognition, in: Proceedings of the IEEE Conference on Computer Vision and Pattern Recognition, 2016, pp. 770–778.

[17] J. Deng, W. Dong, R. Socher, L. Li, Kai Li, Li Fei-Fei, Imagenet: a large-scale hierarchical image database, in: 2009 IEEE Conference on Computer Vision and Pattern Recognition, 2009, pp. 248–255.

[18] R. Doyle, R. Some, W. Powell, G. Mounce, M. Goforth, S. Horan, M. Lowry, High performance spaceflight computing (HPSC) next-generation space processor (NGSP): a joint investment of NASA and AFRL, in: Proceedings of the Workshop on Spacecraft Flight Software, 2013.

[19] R.M. Swan, D. Atha, H.A. Leopold, M. Gildner, S. Oij, C. Chiu, M. Ono, AI4MARS: a dataset for terrain-aware autonomous driving on Mars, in: Proceedings of the IEEE/CVF Conference on Computer Vision and Pattern Recognition (CVPR) Workshops, IEEE/CVF, 2021, p. 5.

[20] R. Arvidson, P. DeGrosse Jr, J. Grotzinger, M. Heverly, J. Shechet, S. Moreland, M. Newby, N. Stein, A. Steffy, F. Zhou, et al., Relating geologic units and mobility system kinematics contributing to curiosity wheel damage at Gale crater, Mars, Journal of Terramechanics 73 (2017) 73–93.

[21] K. He, G. Gkioxari, P. Dollár, R.B. Girshick, Mask R-CNN, CoRR abs/1703.06870, arXiv:1703.06870.
[22] S. Higa, Y. Iwashita, K. Otsu, M. Ono, O. Lamarre, A. Didier, M. Hoffmann, Vision-based estimation of driving energy for planetary rovers using deep learning and terramechanics, IEEE Robotics and Automation Letters 4 (4) (2019) 3876–3883, https://doi.org/10.1109/LRA.2019.2928765.
[23] C. Liu, B. Zoph, M. Neumann, J. Shlens, W. Hua, L.-J. Li, L. Fei-Fei, A. Yuille, J. Huang, K. Murphy, Progressive neural architecture search, in: Proceedings of the European Conference on Computer Vision (ECCV), 2018.
[24] K. Xu, J.L. Ba, R. Kiros, K. Cho, A. Courville, R. Salakhutdinov, R.S. Zemel, Y. Bengio, Show, attend and tell: neural image caption generation with visual attention, in: 32nd International Conference on Machine Learning (ICML), 2015.
[25] G. Hinton, N. Srivastava, K. Swersky, Neural networks for machine learning lecture 6a overview of mini-batch gradient descent, 2012.
[26] D.P. Kingma, J. Ba, Adam: a method for stochastic optimization, arXiv:1412.6980 [cs.LG], 2014.
[27] J. Deng, W. Dong, R. Socher, L. Li, Kai Li, Li Fei-Fei, Bleu: a method for automatic evaluation of machine translation, in: Proceedings of the 40th Annual Meeting of the Association for Computational Linguistics (ACL), 2002.
[28] D.S. Bayard, D.T. Conway, R. Brockers, J.H. Delaune, L.H. Matthies, H.F. Grip, G.B. Merewether, T.L. Brown, A.M.S. Martin, Vision-based navigation for the NASA Mars helicopter, https://doi.org/10.2514/6.2019-1411.

CHAPTER 10

Combining machine-learned regression models with Bayesian inference to interpret remote sensing data

Saverio Cambioni[a], Erik Asphaug[b], and Roberto Furfaro[c]
[a]Department of Earth, Atmospheric & Planetary Sciences, Massachusetts Institute of Technology, Cambridge, MA, United States
[b]Lunar and Planetary Laboratory, University of Arizona, Tucson, AZ, United States
[c]Systems and Industrial Engineering Department, University of Arizona, Tucson, AZ, United States

10.1 The need for accurate fast-forward functions

Increasing computational power and the development of better simulation software with improved physics have allowed the development of models of planetary science phenomena that are much more realistic and include more details than in the past. While these advances shrink the discrepancy between our observations of physical systems and the way we model them, a single model evaluation (simulation or numerical experiment) may require minutes, hours or even days of computing time, and thousands or even millions of model evaluations may be required to understand which model parameters best fit the data. This depends on the number of model parameters (dimensionality), the spatial, temporal and spectral resolution of the data, and the accuracy that is needed. In this review we describe how to use supervised machine learning to design approximation functions called surrogate models and to use them to infer physical properties of asteroids. Surrogate models mimic the behavior of model evaluations (full simulations) within a known level of accuracy, while being computationally orders of magnitude cheaper to evaluate. This allows using surrogate models in lieu of the more computationally expensive parent models to efficiently retrieve the properties of planetary bodies.

10.2 Bayesian approach to inverse problems

An inverse problem aims to retrieve the properties of a physical system by interpreting observations of such system. The system is modeled as $\boldsymbol{y} =$

Machine Learning for Planetary Science
https://doi.org/10.1016/B978-0-12-818721-0.00020-3
Copyright © 2022 Elsevier Inc.
All rights reserved.

$f(\boldsymbol{x}, \boldsymbol{p})$, where $\boldsymbol{y} \in \mathbb{R}^M$ is the output associated with the free parameters $\boldsymbol{x} \in \mathbb{R}^N$ and known parameters $\boldsymbol{p} \in \mathbb{R}^S$. We collected some noisy measurements $\boldsymbol{m} \in \mathbb{R}^M$ of the output $\boldsymbol{y}$ and we want to understand which combination of free parameters $\boldsymbol{x}$ is more likely to correspond to such observations (best-fit $\boldsymbol{x}$).

The modern approach to inverse problems is to treat the free parameters $\boldsymbol{x}$ as random variables (see [47], for a review). The goal is therefore not to find a single best-fit value of $\boldsymbol{x}$, but to estimate the likelihood of all possible values of $\boldsymbol{x}$ to explain the observations $\boldsymbol{m}$. This probability distribution $\pi(\boldsymbol{x}|\boldsymbol{m})$ is commonly referred to as *a-posteriori* or posterior distribution given a set of observations $\boldsymbol{m}$. Since all variables are assumed to be random, they reflect the uncertainty of the true values and, importantly, the degree of uncertainty is encoded in the probability distribution of the parameters to be retrieved. The distribution $\pi(\boldsymbol{x}|\boldsymbol{m})$ is obtained by virtue of Bayes' theorem [4,68]:

$$\pi(\boldsymbol{x}|\boldsymbol{m}) = \frac{\pi_{pr}(\boldsymbol{x})L(\boldsymbol{m}|\boldsymbol{x})}{\pi(\boldsymbol{m})} \tag{10.1}$$

where the probability distribution $\pi_{pr}(\boldsymbol{x})$ encodes prior knowledge or belief about the free parameters, the distribution $L(\boldsymbol{m}|\boldsymbol{x})$ describes the probability of retrieving the observations $\boldsymbol{m}$ given a set of free parameters $\boldsymbol{x}$, and $\pi(\boldsymbol{m})$ is the probability of observing the data $\boldsymbol{m}$:

$$\pi(\boldsymbol{m}) = \int_x \pi(\boldsymbol{x}, \boldsymbol{m})\ dx = \int_x \pi_{pr}(x)L(\boldsymbol{m}|\boldsymbol{x})\ dx \tag{10.2}$$

For large dimensionality N of the free parameters $\boldsymbol{x}$, traditional numerical quadrature schemes for the solution of Eq. (10.2) are not practical and one has to resort to Monte Carlo (MC) integration methods (see [41], for a review). An efficient approach to sampling is to adopt a Markov Chain Monte Carlo (MCMC) scheme and update the sampling using algorithms such as the Delayed Rejection (DR) Adaptive Metropolis-Hastings (AM) (DRAM, [42]).

The computational cost associated with solving Eq. (10.2), however, also depends on the runtime of the model f, as the MCMC scheme requires sampling the unknown $\pi(\boldsymbol{x}|\boldsymbol{m})$ thousands of times. It follows that the use of a faster, but still accurate, version of the model f (a surrogate model) may improve the tractability of the inverse problem, as we shall see in the following Section 10.3.

10.3 Machine-learning-based surrogate models

Building surrogate models is a data-driven task if it occurs without underlying assumptions or a priori knowledge about the model. To this end, among the various techniques of surrogate modeling (see [48], for a review), here we discuss the use of supervised machine learning to train neural networks on a collection of planetary science model simulations. We refer to the computationally intensive physical model as the "parent model", while the "surrogate model" is the neural network trained on the dataset of simulations built using the parent model.

The advantage of using machine learning to build surrogate models resides in its capability to generalize from a sparse look-up table of data in a high-dimensional parameter space. Depending on the nature of the parent model, building a data set to train a neural network is computationally more tractable than finely mapping the parameter space of possible solutions using the parent model directly. On the other hand, neural networks are not naturally protected against overfitting the data (unless care is taken to prevent it). Protection of the networks against overfitting means preventing them to be fooled by chance occurrences in data that do look like interesting patterns but that in fact do not generalize to new unseen data [60]. Furthermore, neural networks do not provide immediate access to conclusions about the underlying physics of the phenomena (causal inference) and cannot do any better than their parent models. This may prevent a clear assessment of the physical consistency of the surrogate model if the researcher lacks the necessary field expertise. Further challenges include errors and biases in the training dataset and suboptimal convergence of the training procedure. Current research efforts to mitigate these issues aim to promote the development of explainable and responsible artificial intelligence (e.g., [1]). The use of machine-learned surrogate models is relatively new in planetary science, but applications here and in astronomy and geophysics show the potential of the method. These include predicting the outcomes of collisions between planets to improve the realism of terrestrial planet formation (e.g., [10,24,25]); predicting the stability of circumbinary planets on initially coplanar, circular orbits [52]; characterizing the interior structure of planets with mass below 25 Earth masses [7]; modeling mantle convection on Earth [5]; quantifying uncertainties in highly nonlinear models such as those used to study seismology measurements [17]; and emulating the physics of general circulation models [65].

10.4 Case study: constraining the thermal properties of asteroids with surrogate models

The field of asteroid thermophysical modeling has experienced an extraordinary growth in the last years, as new thermal infrared data became available for hundreds of thousands of asteroids [19]. Here we describe how to use machine learning to design a surrogate thermophysical model for asteroids and to employ it to retrieve the surface roughness, the thermal inertia of the rocks and unconsolidated material (regolith) on the surface, and the relative abundances of rocky material and regolith. The thermal inertia Γ measures a material's resistance to change temperature with time. It is defined as $\Gamma = \sqrt{k\rho c}$, where k is the thermal conductivity, ρ is the bulk density, and c is the mean specific heat. It has units of $\mathrm{J\,s^{-1/2}\,m^{-2}\,K^{-1}}$. The case study discussed here is published in its entirety in [12], where the surrogate thermophysical model is used to characterize the thermal properties of near-Earth asteroid 25143 Itokawa (Fig. 10.1) from an ensemble of mid-infrared observations of the surface. [12] interpreted the results in the context of the observations by JAXA Hayabusa mission, which visited Itokawa between November 2005 and December 2007 and characterized its surface in great detail [50].

10.4.1 Dataset of thermophysical simulations

In [12], the parent model $\boldsymbol{y} = f(\boldsymbol{x}, \boldsymbol{p})$ is the thermophysical model of asteroid regolith by [19]. The free parameters $\boldsymbol{x}$ are the thermal inertia Γ of the surface material and the roughness of the surface θ. The known parameters $\boldsymbol{p}$ are the shape model of Itokawa [36], the asteroids' position and orientation in space with respect to the observer, the illumination condition at each time of observation and the wavelength of the observation. The output $\boldsymbol{y}$ is the infrared flux from the surface corresponding to a certain set $\boldsymbol{x} = (\Gamma, \theta)$ for a given $\boldsymbol{p}$. Using the ThermoPhysical Model by [19] a look-up table of simulated infrared fluxes is built for all the possible combinations of (Γ, θ) in the following range:

- $\Gamma \in [25, 2500]$ with step size $25\,\mathrm{J\,s^{-1/2}\,m^{-2}\,K^{-1}}$;
- $\theta \in [0, 60]$ with step size $10°$.

The thermal inertia value ranges between $25\,\mathrm{J\,s^{-1/2}\,m^{-2}\,K^{-1}}$ (similar to that of fine regolith on the surface of the Moon and asteroid Vesta, [14,44]) to $2500\,\mathrm{J\,s^{-1/2}\,m^{-2}\,K^{-1}}$ (similar to low-porosity basalt, [71]) with a step of $25\,\mathrm{J\,s^{-1/2}\,m^{-2}\,K^{-1}}$. For each of the 30 infrared observations analyzed in [12], 700 simulations are therefore run. This means that, in total, the final

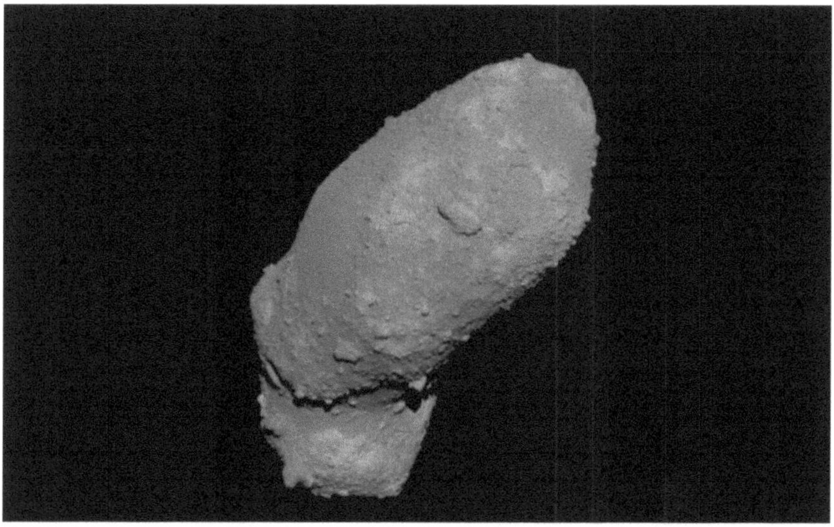

Figure 10.1 Image of asteroid (25143) Itokawa acquired on October 10, 2005 by the AMICA instrument on JAXA Hayabusa mission. This image (HAY-A-AMICA-3-HAYAMICA-V1.0_20051015_ST_2461722826_V_FIT) was obtained from the NASA Planetary Data System as part of the Hayabusa AMICA data set [62]. The image has been re-processed using the SAOImageDS9 software [49] for visualization purposes.

dataset of thermophysical simulations contains 21,000 simulations (i.e., 700 simulations times 30 observations). Building such a dataset requires a total run time of 5 days on a single processor.

10.4.2 Surrogate infrared model of a mixture of regolith and rock

The surface of asteroids, however, is more accurately modeled by a mixture of both subcentimeter regolith and rocks, as also suggested by the presence of both smooth and rough terrains observed by the Hayabusa mission on the surface of Itokawa (Fig. 10.1). On asteroids, regolith smaller than a few centimeters in diameter is produced in the break-up of larger rocks by micrometeoroid impacts [6,45,46] and thermal fatigue cracking, caused by the intense temperature variations between day and night [18,22,55]. Subcentimeter regolith particulate on asteroids has a lower thermal inertia than solid rocks of the same composition [19]. Following this, in the dataset of simulations of Section 10.4.1, the simulations with thermal inertia values Γ_R greater than 700 $\mathrm{J\,s^{-1/2}\,m^{-2}\,K^{-1}}$ (i.e., the lowest value recorded for stony meteorites, [57]) are assumed to model the infrared emission by rocks. Simulations with thermal inertia Γ_P lower than $700\,\mathrm{J\,s^{-1/2}\,m^{-2}\,K^{-1}}$ are assumed to model the infrared emission by unconsolidated (but not necessarily homogeneous) particulate, i.e., the regolith. The rock and regolith units linearly contribute to the emitted infrared flux with a weight equal to the areal abundance of rock (termed rock abundance, or *RA* hereafter) relative to that of regolith. For each observation of Itokawa's surface, the results (infrared flux f) of the simulations of Section 10.4.1 corresponding to $\Gamma < 700$ $\mathrm{J\,s^{-1/2}\,m^{-2}\,K^{-1}}$ (regolith) and those with $\Gamma > 700\,\mathrm{J\,s^{-1/2}\,m^{-2}\,K^{-1}}$ (rock) are linearly combined as:

$$\boldsymbol{F}(\theta, \Gamma_R, \Gamma_P, RA) = f_{Rock}(\theta, \Gamma_R)\; RA + f_{Regolith}(\theta, \Gamma_P)\; (1 - RA) \qquad (10.3)$$

where the rock abundance *RA* ranges from 0–100% with step size 5%. This results in a dataset with entries:

$$\{\boldsymbol{x}, \boldsymbol{y}\} = \{(\theta, \Gamma_R, \Gamma_P, RA); \boldsymbol{F}(\theta, \Gamma_R, \Gamma_P, RA)\} \qquad (10.4)$$

For each observation of Itokawa's surface listed in Table 3 in [12], the correspondent dataset described by Eq. (10.4) is used to train a neural network (thermal surrogate model, Fig. 10.2). Following hyperparameter optimization, each network has a single hidden layer with 10 neurons and is trained using the Levenberg-Marquardt algorithm [20] with 10-fold cross-validation. The MATLAB® Neural Network package is used to perform the training on 70% of the dataset by backpropagating the residuals between the predicted thermal flux and the simulated thermal flux for all samples in the training dataset (see Chapter 1 for a more detailed explanation of backpropagation in neural networks). The goal of the training procedure is to find the parameters of the network (weights or coefficients $\boldsymbol{w}$) that minimize the Mean Squared Error:

$$\epsilon(\boldsymbol{w}) = \frac{1}{K}\sum_{i=1}^{K}[\boldsymbol{F}_{NN}(\theta, \Gamma_R, \Gamma_P, RA) - \boldsymbol{F}(\theta, \Gamma_R, \Gamma_P, RA)]^2 \qquad (10.5)$$

where F_{NN} is the infrared flux predicted by the neural network and K is the number of samples in each training batch. In this respect, the training procedure is not a simple interpolation of the data, but it is rather a search for classes of parametric functions that globally fit the data (i.e., the neural network that best generalizes the input-output relationship underlying the thermophysical parent model). This is done by using the validation set (15% of the dataset) to stop training before the model begins to overfit the training data, as evidenced by a drop in validation performance, and to assess the performance of the trained network, respectively [9]. The success in achieving optimal generalization capabilities depends on how much correlation exists between the training and validation datasets. In this case study, the datasets are composed by data that come from different, independent simulations. In general, we caution readers to ensure independence as much as possible between samples in their training, validation, and test sets. The network's performance is evaluated using the Mean Squared Error and correlation index on the testing set, which is composed by 15% of data that were not previously used for training or validation. For a summary of the performances of the networks in [12], we refer the reader to Table 5 in that work.

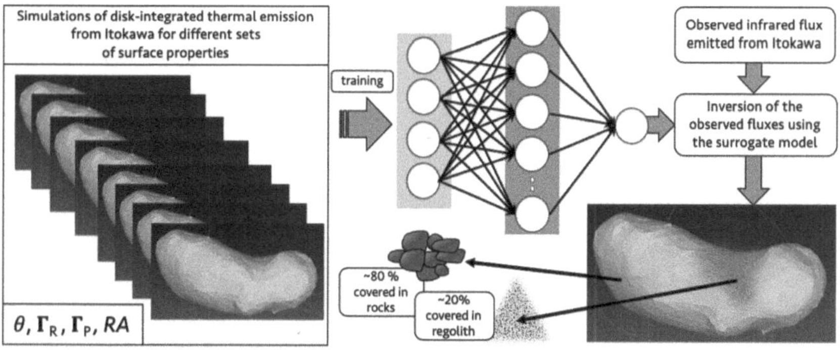

Figure 10.2 Flowchart of the procedure by [12] to derive the surface roughness θ, the thermal inertia of the rock Γ_R, the thermal inertia of the regolith particulate Γ_P and the rock abundance *RA* on Itokawa's surface from measurements of the emitted infrared flux. From left to right: the dataset of thermophysical simulations of Section 10.4.1 is used to train a neural network that predicts the infrared fluxes correspondent to a specific set of surface properties (θ, Γ_R, Γ_P, *RA*). For each observation of Itokawa's surface, a neural network is then used in the Bayesian inversion of the observed infrared flux of Itokawa's surface to infer its properties. Itokawa's shape model is from [36].

10.4.3 Bayesian inference of Itokawa's thermophysical properties

This section summarizes the findings by [12] using the neural networks of Section 10.4.2 as the forward model in the Bayesian inversion described in Section 10.2. The inversion is supported by the following a priori information that defines the distribution $\pi_{pr}(\boldsymbol{x})$:

- Regolith on Itokawa can be treated as an homogeneous material as long as the particle size does not exceed the penetration depth of the diurnal temperature cycle (about 2 cm). Laboratory experiments (e.g., [63]) allow converting the grain size of the regolith in thermal conductivity under reasonable assumptions for the density, porosity and thermal capacity of the grain. The condition of homogeneity of the regolith material yields an upper limit for the range of solutions in regolith thermal inertia equal to 285 $\mathrm{J\,s^{-1/2}\,m^{-2}\,K^{-1}}$.

- The size frequency distribution for boulders larger than 5 m on the surface of Itokawa from Hayabusa/AMICA images [54] yields a lower limit for the rock abundance equal to 16%.

Fig. 10.3 plots the probability distributions $\pi\{(\theta, \Gamma_{regolith}, \Gamma_{rock}, RA)|\boldsymbol{m}\}$, which show the likelihood of a certain set of thermal properties to be representative of Itokawa's surface given the thermal infrared observations listed in Table 3 of [12]. The results of the Bayesian inversion indicate that:

- The surface of asteroid Itokawa is dominated by rocks (RA = 84 ±9%). This is consistent with observations by the Hayabusa mission which showed that only 20% of the surface of Itokawa is covered in subcentimeter regolith [61,69].
- The thermal inertia of the rocks on Itokawa is estimated to be $\Gamma_R = 894 \pm 122\,\mathrm{J\,s^{-1/2}\,m^{-2}\,K^{-1}}$. This value is lower than the thermal inertia of the rock measured value for LL chondritic meteorites [57], suggesting that the rocks on Itokawa could be fractured.
- The thermal inertia of the regolith is $\Gamma_P = 203 \pm 36\ \mathrm{J\,s^{-1/2}\,m^{-2}\,K^{-1}}$. Assuming macroporosity between 40% and 60%, the best-fit value of regolith thermal inertia suggests an average particle size of about 1 cm according to the method by [63]. This is also in agreement with in situ findings by Hayabusa [51].

10.5 Future perspectives for data fusion

We conclude describing some research directions that will benefit from the use of surrogate models in Bayesian inversion of observed data.

10.5.1 Remote sensing data fusion

The recent advancements in machine learning and Bayesian inversion allow not only more efficient and accurate interpretation of planetary science data, but also the joint-inversion of measurements collected by different instruments on telescopes, spacecrafts and laboratory experiments. This approach is typically referred to as data fusion [15].

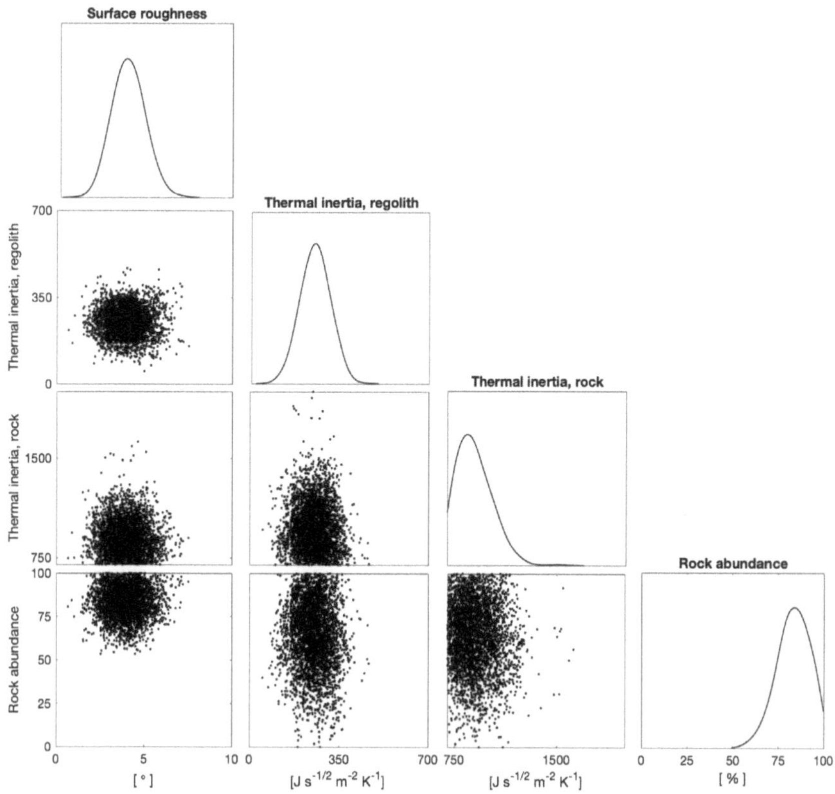

Figure 10.3 The posterior distributions of the thermophysical properties of the surface of Itokawa from the Bayesian inversion of observed infrared fluxes indicate that the surface is covered with about 20% in regolith smaller than a few centimeters (consistent with observations by the Hayabusa mission, [61,69]). The thermal inertia of the rocks indicates that most of the rocks could be fractured [12].

An example is the data fusion of remotely-sensed radar echo power and surface infrared emission from planetary icy surfaces [11]. This approach allows retrieving the temperature profile and average composition of the ice *independently* from one another, thus disentangling the effect of ice temperature and presence of impurities on the radar echo without extracting ice cores (for methods based on ice core data, see [53,70]). While this data fusion method is relevant to the characterization of terrestrial ice sheets (e.g., the case of Lake Vostok in East Antarctica, [11]), we foresee that it will be essential to maximize the scientific return of the forthcoming NASA Clipper and ESA JUICE missions to Jupiter's moon Europa. Both missions will carry a radar sounder and a thermal imager, but will not drill into the Europan ice.

The data fusion method by [11] also estimates the surface porosity and densification of the ice that is crucial knowledge for designing future landed missions on icy satellites. Future landers (e.g., Europa lander, [58]) will also be equipped with seismometers that can collect the seismic signature of the interaction between the rocky ocean floor and the water [13] where biological reactions can conceivably occur [43]. Radar sounding and seismic measurements are highly complementary probes of the upper kilometers of ice. Future research directions include investigating whether the fusion of the three datasets (thermal, radar and seismic data) enables a full mapping of Europa's interior structure and geological activity, and where seismographic measurements should be collected on the surface in order to resolve ambiguities in the subsurface structure and ongoing and past geological activity.

10.5.2 Planet formation theory

Surrogate models also represent a significant advancement in planet formation theory, where we have the riddle of the incredibly diverse bodies which compose the Main Belt (e.g., asteroid 16 Psyche), the Kuiper Belt, and the inner solar system planetary population (e.g., planet Mercury, [2]).

For Mercury, early studies proposed its catastrophic disruption by a projectile [8] to explain the unusually high core mass fraction. However, the 2011-2012 NASA MESSENGER mission showed that the planet has an anomalously high volatile content compared to the Moon [59]. On this basis, [3] argued that two lower-intensity collisions in a row would be a better solution. These data – including Mercury's reduced oxidation state compared to the other planets [56] and hints for Mercury's meteorites in our collections (e.g., [21]) – can be fused into the problem of Mercury's collision history. As noted, the surrogate model is a function, which means that for given astronomical observations, spacecraft remote sensing and laboratory measurements a Bayesian inversion can be performed to obtain the most likely impact hypothesis.

Data for asteroid 16 Psyche are still sparse [23], but this will change when the eponymous NASA mission arrives in 2026, carrying a color imager, a magnetometer to characterize the metallic core, and a gamma-ray and neutron spectrometer to measure composition. Presently there is much confusion as to whether Psyche is, in fact, a planetary core, as bulk density estimates range from around 2000 kg/m^3 [67] to more than 4000 kg/m^3 [26]. When bulk density and composition are known, and the extent of the core, these stronger constraints can be applied to obtain a collision history. Psyche may have lost its mantle in multiple collisions, as Mercury probably did [2]. Surrogate models that are trained on this complicated collisional regime will be essential to interpret all the available datasets.

10.5.3 Spacecraft autonomy

Finally, machine learning and surrogate modeling will play a major role in enabling high-level spacecraft autonomy, which is a critical feature for future science-driven and unconstrained missions on solar system bodies. To expand the knowledge of such bodies, one will need the design, development and deployment of a new class of autonomous agents (e.g., orbiters, landers, rovers, balloons, UAVs) that enable intelligent planetary reconnaissance [27,28,35]. Such systems could include artificial intelligence-driven software that enables fully automated identification and characterization of sites that yield the highest potential for scientific discovery. This will allow integrating ancillary information with spatial and temporal sensor data acquired during the course of planetary exploration, thus enabling reasoning and autonomous deployment and site selection [33].

Over the past few years, there has been a surge in researching how deep learning methods can be applied to the problem of autonomous, closed-loop guidance, navigation and control for spacecraft landing on planetary bodies. For example, supervised deep learning has been explored as possible mechanisms for closed-loop guidance for landing problems [30,31,64]. Deep reinforcement learning, where the spacecraft agent interacts with the dynamical environment and learns a specific task by maximizing a reward, has been applied for guidance and control of planetary landing [37,38,66] and asteroid close-proximity operations [39,40]. Recent effort in surrogate modeling of the gravitational field of irregular small bodies may enable real-time, on-board generation of optimal spacecraft trajectories near asteroids and comets [16,29]. We foresee these methodologies to be critical to implement the level of autonomy necessary to sustain the next wave of space exploration in the solar system and to maximize the potential for new discoveries in the search for life (e.g., [32]) and characterization of planetary features (e.g., [34]).

Acknowledgments. S.C., E.A. and R.F. acknowledge support from NASA under grant 80NSSC19K0817. The authors thank the anonymous referees for the precious comments and suggestions that improved this review.

Data availability. The data that support the findings published in [12] and described in this review are available from the corresponding author upon reasonable request.

References

[1] A.B. Arrieta, N. Díaz-Rodríguez, J. Del Ser, A. Bennetot, S. Tabik, A. Barbado, S. García, S. Gil-López, D. Molina, R. Benjamins, et al., Explainable Artificial Intelligence (XAI): concepts, taxonomies, opportunities and challenges toward responsible AI, Information Fusion 58 (2020) 82–115.

[2] E. Asphaug, Signatures of hit-and-run collisions, in: L.T. Elkins-Tanton, B.P. Weiss (Eds.), Planetesimals: Early Differentiation and Consequences for Planets. Cambridge Planetary Science, Cambridge University Press, 2017, pp. 7–37.

[3] E. Asphaug, A. Reufer, Mercury and other iron-rich planetary bodies as relics of inefficient accretion, Nature Geoscience 7 (Aug. 2014) 564–568.

[4] R.C. Aster, B. Borchers, C.H. Thurber, Parameter Estimation and Inverse Problems, vol. 90, Academic Press, 2011.

[5] S. Atkins, A.P. Valentine, P.J. Tackley, J. Trampert, Using pattern recognition to infer parameters governing mantle convection, Physics of the Earth and Planetary Interiors 257 (Aug. 2016) 171–186.

[6] A.T. Basilevsky, J.W. Head, F. Horz, Survival times of meter-sized boulders on the surface of the Moon, Planetary and Space Science 89 (Dec. 2013) 118–126.

[7] P. Baumeister, S. Padovan, N. Tosi, G. Montavon, N. Nettelmann, J. MacKenzie, M. Godolt, Machine-learning inference of the interior structure of low-mass exoplanets, The Astrophysical Journal 889 (1) (Jan. 2020) 42.

[8] W. Benz, A. Anic, J. Horner, J.A. Whitby, The origin of Mercury, Space Science Reviews 132 (Oct. 2007) 189–202.

[9] C.M. Bishop, et al., Neural Networks for Pattern Recognition, Oxford University Press, 1995.

[10] S. Cambioni, E. Asphaug, A. Emsenhuber, T.S.J. Gabriel, R. Furfaro, S.R. Schwartz, Realistic on-the-fly outcomes of planetary collisions: machine learning applied to simulations of giant impacts, The Astrophysical Journal 875 (1) (2019) 40.

[11] S. Cambioni, L. Carter, M. Haynes, E. Asphaug, R. Furfaro, Machine learning for characterizing shallow subsurface ice via radar-thermal data fusion: validation at Lake Vostok, East Antarctica, in: European Planetary Science Congress—Division of Planetary Science Joint Meeting 2019, vol. 13, 2019, p. 192-1.

[12] S. Cambioni, M. Delbo, A.J. Ryan, R. Furfaro, E. Asphaug, Constraining the thermal properties of planetary surfaces using machine learning: application to airless bodies, Icarus 325 (2019) 16–30.

[13] F. Cammarano, V. Lekic, M. Manga, M. Panning, B. Romanowicz, Long-period seismology on Europa: 1. Physically consistent interior models, Journal of Geophysical Research: Planets 111 (E12) (2006).

[14] M.T. Capria, F. Tosi, M. De Sanctis, F. Capaccioni, E. Ammannito, A. Frigeri, F. Zambon, S. Fonte, E. Palomba, D. Turrini, et al., Vesta surface thermal properties map, Geophysical Research Letters 41 (5) (2014) 1438–1443.

[15] N.-B. Chang, K. Bai, Multisensor Data Fusion and Machine Learning for Environmental Remote Sensing, CRC Press, 2018.

[16] L. Cheng, Z. Wang, Y. Song, F. Jiang, Real-time optimal control for irregular asteroid landings using deep neural networks, Acta Astronautica 170 (2020) 66–79.

[17] R.W.L. de Wit, A.P. Valentine, J. Trampert, Bayesian inference of Earth's radial seismic structure from body-wave traveltimes using neural networks, Geophysical Journal International 195 (1) (Oct. 2013) 408–422.

[18] M. Delbo, G. Libourel, J. Wilkerson, N. Murdoch, P. Michel, K.T. Ramesh, C. Ganino, C. Verati, S. Marchi, Thermal fatigue as the origin of regolith on small asteroids, Nature 508 (7) (Apr. 2014) 233–236.
[19] M. Delbo, M. Mueller, J.P. Emery, B. Rozitis, M.T. Capria, Asteroid thermophysical modeling, in: P. Michel, et al. (Eds.), Asteroids IV, University of Arizona Press, Tucson, 2015, pp. 107–128.
[20] H.B. Demuth, M.H. Beale, O. De Jess, M.T. Hagan, Neural network design. Martin Hagan, 2014.
[21] D.S. Ebel, S.T. Stewart, The elusive origin of Mercury, in: S. Solomon, L.R. Nittler, B.J. Anderson (Eds.), Mercury: The View After MESSENGER, vol. 21, Cambridge University Press, 2018, pp. 497–515.
[22] C. El Mir, K. Hazeli, K.T. Ramesh, M. Delbo, J. Wilkerson, Thermal fatigue: length-scales, timescales, and their implications on regolith size-frequency distribution, in: 47th Lunar and Planetary Science Conference, vol. 47, Mar. 2016, p. 2586.
[23] L. Elkins-Tanton, E. Asphaug, J. Bell III, H. Bercovici, B. Bills, R. Binzel, W. Bottke, S. Dibb, D. Lawrence, S. Marchi, et al., Observations, meteorites, and models: a preflight assessment of the composition and formation of (16) Psyche, Journal of Geophysical Research: Planets 125 (3) (2020) e2019JE006296.
[24] A. Emsenhuber, S. Cambioni, collresolve, Dec. 2019.
[25] A. Emsenhuber, S. Cambioni, E. Asphaug, T.S.J. Gabriel, S.R. Schwartz, R. Furfaro, Realistic on-the-fly outcomes of planetary collisions. II. Bringing machine learning to N-body simulations, The Astrophysical Journal 891 (1) (Mar. 2020) 6.
[26] M. Ferrais, P. Vernazza, L. Jorda, N. Rambaux, J. Hanus, Asteroid (16) psyche's primordial shape: a possible Jacobi ellipsoid, Astronomy & Astrophysics (2020).
[27] W. Fink, J.M. Dohm, M.A. Tarbell, T.M. Hare, V.R. Baker, Next-generation robotic planetary reconnaissance missions: a paradigm shift, Planetary and Space Science 53 (14–15) (2005) 1419–1426.
[28] W. Fink, J.M. Dohm, M.A. Tarbell, T.M. Hare, V.R. Baker, D. Schulze-Makuch, R. Furfaro, A.G. Fairén, T.P. Ferré, H. Miyamoto, et al., Tier-scalable reconnaissance missions for the autonomous exploration of planetary bodies, in: 2007 IEEE Aerospace Conference, IEEE, 2007, pp. 1–10.
[29] R. Furfaro, R. Barocco, R. Linares, T. Francesco, V. Reddy, J. Simo, L. Le Corre, Modeling irregular small bodies gravity field via extreme learning machines and Bayesian optimization, Advances in Space Research (2020).
[30] R. Furfaro, I. Bloise, M. Orlandelli, P. Di Lizia, F. Topputo, R. Linares, et al., Deep learning for autonomous lunar landing, in: 2018 AAS/AIAA Astrodynamics Specialist Conference, 2018, pp. 1–22.
[31] R. Furfaro, I. Bloise, M. Orlandelli, P. Di Lizia, F. Topputo, R. Linares, et al., A recurrent deep architecture for quasi-optimal feedback guidance in planetary landing, in: IAA SciTech Forum on Space Flight Mechanics and Space Structures and Materials, 2018, pp. 1–24.
[32] R. Furfaro, J. Dohm, W. Fink, J. Kargel, D. Schulze-Makuch, A. Fairén, A. Palmero-Rodriguez, V. Baker, P. Ferré, T. Hare, et al., The search for life beyond Earth through fuzzy expert systems, Planetary and Space Science 56 (3–4) (2008) 448–472.
[33] R. Furfaro, W. Fink, J.S. Kargel, Autonomous real-time landing site selection for Venus and Titan using evolutionary fuzzy cognitive maps, Applied Soft Computing 12 (12) (2012) 3825–3839.
[34] R. Furfaro, J.S. Kargel, J.I. Lunine, W. Fink, M.P. Bishop, Identification of cryovolcanism on Titan using fuzzy cognitive maps, Planetary and Space Science 58 (5) (2010) 761–779.
[35] R. Furfaro, J.I. Lunine, J.S. Kargel, W. Fink, Intelligent systems for the autonomous exploration of Titan and Enceladus, in: Space Exploration Technologies, vol. 6960, International Society for Optics and Photonics, 2008, p. 69600A.

[36] R. Gaskell, J. Saito, M. Ishiguro, T. Kubota, T. Hashimoto, N. Hirata, S. Abe, O. Barnouin-Jha, D. Scheeres, Gaskell Itokawa Shape Model V1.0. NASA Planetary Data System 92, HAY–A–AMICA–5–ITOKAWASHAPE–V1.0, Sep. 2008.
[37] B. Gaudet, R. Linares, R. Furfaro, Adaptive guidance and integrated navigation with reinforcement meta-learning, Acta Astronautica 169 (2020) 180–190.
[38] B. Gaudet, R. Linares, R. Furfaro, Deep reinforcement learning for six degree-of-freedom planetary landing, Advances in Space Research 65 (7) (2020) 1723–1741.
[39] B. Gaudet, R. Linares, R. Furfaro, Six degree-of-freedom body-fixed hovering over unmapped asteroids via lidar altimetry and reinforcement meta-learning, Acta Astronautica 172 (2020) 90–99.
[40] B. Gaudet, R. Linares, R. Furfaro, Terminal adaptive guidance via reinforcement meta-learning: applications to autonomous asteroid close-proximity operations, Acta Astronautica 171 (2020) 1–13.
[41] L.A. Glasgow, Applied Mathematics for Science and Engineering, Wiley Online Library, 2014.
[42] H. Haario, M. Laine, A. Mira, E. Saksman, DRAM: efficient adaptive MCMC, Statistics and Computing 16 (4) (2006) 451–559.
[43] K.P. Hand, C.F. Chyba, J.C. Priscu, R.W. Carlson, K.H. Nealson, Astrobiology and the potential for life on Europa, in: Robert T. Pappalardo, William B. McKinnon, Krishan K. Khurana (Eds.), Europa, University of Arizona Press, Tucson, 2009, p. 589.
[44] P.O. Hayne, J.L. Bandfield, M.A. Siegler, A.R. Vasavada, R.R. Ghent, J.-P. Williams, B.T. Greenhagen, O. Aharonson, C.M. Elder, P.G. Lucey, et al., Global regolith thermophysical properties of the Moon from the diviner lunar radiometer experiment, Journal of Geophysical Research: Planets 122 (12) (2017) 2371–2400.
[45] F. Hoerz, E. Schneider, D.E. Gault, J.B. Hartung, D.E. Brownlee, Catastrophic rupture of lunar rocks - a Monte Carlo simulation, Lunar Science Institute 13 (1–3) (Jul. 1975) 235–258.
[46] F. Hörz, M. Cintala, Impact experiments related to the evolution of planetary regoliths, Meteoritics & Planetary Science 32 (2) (Mar. 1997) 179–209.
[47] J. Idier, Bayesian Approach to Inverse Problems, John Wiley & Sons, 2013.
[48] P. Jiang, Q. Zhou, X. Shao, Surrogate Model-Based Engineering Design and Optimization, Springer, 2020.
[49] W.A. Joye, E. Mandel, New features of SAOImage DS9, in: H.E. Payne, R.I. Jedrzejewski, R.N. Hook (Eds.), Astronomical Data Analysis Software and Systems XII, in: Astronomical Society of the Pacific Conference Series, vol. 295, Jan. 2003, p. 489.
[50] J. Kawaguchi, A. Fujiwara, T. Uesugi, Hayabusa—its technology and science accomplishment summary and Hayabusa-2, Acta Astronautica 62 (10) (May 2008) 639–647.
[51] K. Kitazato, B.E. Clark, M. Abe, S. Abe, Y. Takagi, T. Hiroi, O.S. Barnouin-Jha, P.A. Abell, S.M. Lederer, F. Vilas, Near-infrared spectrophotometry of Asteroid 25143 Itokawa from NIRS on the Hayabusa spacecraft, Icarus 194 (1) (Mar. 2008) 137–145.
[52] C. Lam, D. Kipping, A machine learns to predict the stability of circumbinary planets, Monthly Notices of the Royal Astronomical Society 476 (4) (Jun. 2018) 5692–5697.
[53] J.A. MacGregor, D.P. Winebrenner, H. Conway, K. Matsuoka, P.A. Mayewski, G.D. Clow, Modeling englacial radar attenuation at Siple Dome, West Antarctica, using ice chemistry and temperature data, Journal of Geophysical Research: Earth Surface 112 (F3) (2007).
[54] S. Mazrouei, M.G. Daly, O.S. Barnouin, C.M. Ernst, I. DeSouza, Block distributions on Itokawa, Icarus 229 (Feb. 2014) 181–189.
[55] J. Molaro, S. Byrne, Thermal stress weathering on Mercury and other airless bodies, in: Lunar and Planetary Science Conference, vol. 1608, 2011, p. 1494.
[56] L.R. Nittler, N.L. Chabot, T.L. Grove, P.N. Peplowski, The chemical composition of Mercury, in: S. Solomon, L.R. Nittler, B.J. Anderson (Eds.), Mercury: The View After MESSENGER, vol. 21, Cambridge University Press, 2018, pp. 30–51.

[57] C.P. Opeil, G.J. Consolmagno, D.J. Safarik, D.T. Britt, Stony meteorite thermal properties and their relationship with meteorite chemical and physical states, Meteoritics and Planetary Science 47 (3) (Mar. 2012) 319–329.
[58] R. Pappalardo, S. Vance, F. Bagenal, B. Bills, D. Blaney, D. Blankenship, W. Brinckerhoff, J. Connerney, K. Hand, T.M. Hoehler, et al., Science potential from a Europa lander, Astrobiology 13 (8) (2013) 740–773.
[59] P.N. Peplowski, L.G. Evans, S.A. Hauck, T.J. McCoy, W.V. Boynton, J.J. Gillis-Davis, D.S. Ebel, J.O. Goldsten, D.K. Hamara, D.J. Lawrence, et al., Radioactive elements on Mercury's surface from MESSENGER: implications for the planet's formation and evolution, Science 333 (6051) (2011) 1850–1852.
[60] F. Provost, T. Fawcett, Data Science for Business: What You Need to Know About Data Mining and Data-Analytic Thinking, O'Reilly Media, Inc., 2013.
[61] J. Saito, H. Miyamoto, R. Nakamura, M. Ishiguro, T. Michikami, A.M. Nakamura, H. Demura, S. Sasaki, N. Hirata, C. Honda, A. Yamamoto, Y. Yokota, T. Fuse, F. Yoshida, D.J. Tholen, R.W. Gaskell, T. Hashimoto, T. Kubota, Y. Higuchi, T. Nakamura, P. Smith, K. Hiraoka, T. Honda, S. Kobayashi, M. Furuya, N. Matsumoto, E. Nemoto, A. Yukishita, K. Kitazato, B. Dermawan, A. Sogame, J. Terazono, C. Shinohara, H. Akiyama, Detailed images of Asteroid 25143 Itokawa from Hayabusa, Science 312 (Jun. 2006) 1341–1344.
[62] J. Saito, T. Nakamura, H. Akiyama, H. Demura, B. Dermawan, M. Furuya, T. Fuse, R. Gaskell, T. Hashimoto, Y. Higuchi, K. Hiraoka, N. Hirata, C. Honda, T. Honda, M. Ishiguro, K. Kitazato, S. Kobayashi, T. Kubota, N. Matsumoto, T. Michikami, H. Miyamoto, A. Nakamura, R. Nakamura, E. Nemoto, S. Sasaki, C. Shinohara, P. Smith, A. Sogame, J. Terazono, D. Tholen, A. Yamamoto, Y. Yokota, F. Yoshida, A. Yukishita, Hayabusa AMICA V1.0. HAY-A-AMICA-3-HAYAMICA-V1.0. NASA Planetary Data System, 2010.
[63] N. Sakatani, K. Ogawa, Y. Iijima, M. Arakawa, R. Honda, S. Tanaka, Thermal conductivity model for powdered materials under vacuum based on experimental studies, AIP Advances 7 (1) (Jan. 2017) 015310.
[64] C. Sánchez-Sánchez, D. Izzo, Real-time optimal control via Deep Neural Networks: study on landing problems, Journal of Guidance, Control, and Dynamics 41 (5) (2018) 1122–1135.
[65] S. Scher, Toward data-driven weather and climate forecasting: approximating a simple general circulation model with deep learning, Geophysical Research Letters 45 (22) (2018) 12–616.
[66] A. Scorsoglio, R. Furfaro, R. Linares, B. Gaudet, Image-based deep reinforcement learning for autonomous lunar landing, in: AIAA Scitech 2020 Forum, 2020, p. 1910.
[67] L. Siltala, M. Granvik, New estimates of the mass and density of asteroid (16) Psyche, in: European Planetary Science Congress—Division of Planetary Science Joint Meeting 2019, vol. 2019, Sep. 2019, p. 1610.
[68] A.M. Stuart, Inverse problems: a Bayesian perspective, Acta Numerica 19 (2010) 451–559.
[69] H.C. Susorney, C.L. Johnson, O.S. Barnouin, M. Daly, J. Seabrook, D.S. Lauretta, E.B. Bierhaus, The global surface roughness of 25143 Itokawa, in: Lunar and Planetary Science Conference, in: Lunar and Planetary Science Conference, vol. 49, Mar. 2018, p. 1066.
[70] D.P. Winebrenner, P.M. Kintner, J.A. MacGregor, New estimates of ice and oxygen fluxes across the entire lid of Lake Vostok from observations of englacial radio wave attenuation, Journal of Geophysical Research: Earth Surface 124 (3) (2019) 795–811.
[71] J.R. Zimbelman, The role of porosity in thermal inertia variations on basaltic lavas, Icarus 68 (2) (Nov. 1986) 366–369.

Index

L

M

Q

R

FSC
www.fsc.org
MIX
Papier | Fördert gute Waldnutzung
FSC® C083411

Kolibri 360 GmbH
Ferdinand-Jühlke-Straße 7
99095 Erfurt, Deutschland
produktsicherheit@kolibri360.de